LABOUR IN THE LABORATORY

MCGILL-QUEEN'S/ASSOCIATED MEDICAL SERVICES
Studies in the History of Medicine, Health, and Society

Series Editors: S.O. Freedman and J.H. Connor

Volumes in this series have financial support from Associated Medical Services, Inc. (AMS). Associated Medical Services Inc. was established in 1936 by Dr Jason Hannah as a pioneer not-for-profit health care organization in Ontario. With the advent of medicare, AMS became a charitable organization supporting innovations in academic medicine and health services, specifically the history of medicine and health care, as well as innovations in health professional education and bioethics.

1 Home Medicine
  The Newfoundland Experience
  John K. Crellin

2 A Long Way from Home
  The Tuberculosis Epidemic among the Inuit
  Pat Sandiford Grygier

3 Labrador Odyssey
  The Journal and Photographs of Eliot Curwen on the Second Voyage of Wilfred Grenfell, 1893
  Edited by Ronald Rompkey

4 Architecture in the Family Way
  Doctors, Houses, and Women, 1870–1900
  Annmarie Adams

5 Local Hospitals in Ancien Régime France
  Rationalization, Resistance, Renewal, 1530–1789
  Daniel Hickey

6 Foisted upon the Government?
  State Responsibilities, Family Obligations, and the Care of the Dependant Aged in Nineteenth-Century Ontario
  Edgar-André Montigny

7 A Young Man's Benefit
  The Independent Order of Odd Fellows and Sickness Insurance in the United States and Canada, 1860–1929
  George Emery and J.C. Herbert Emery

8 The Weariness, the Fever, and the Fret
  The Campaign against Tuberculosis in Canada, 1900–1950
  Katherine McQuaig

9 The War Diary of Clare Gass, 1915–1918
  Edited by Susan Mann

10 Committed to the State Asylum
   Insanity and Society in Nineteenth-Century Quebec and Ontario
   James E. Moran

11 Jessie Luther at the Grenfell Mission
   Edited by Ronald Rompkey

12 Negotiating Disease
   Power and Cancer Care,
   1900–1950
   Barbara Clow

13 For Patients of Moderate Means
   A Social History of the Voluntary
   Public General Hospital in
   Canada, 1890–1950
   *David Gagan
   and Rosemary Gagan*

14 Into the House of Old
   A History of Residential
   Care in British Columbia
   Megan J. Davies

15 St Mary's
   The History of a London
   Teaching Hospital
   E.A. Heaman

16 Women, Health, and Nation
   Canada and the United States
   since 1945
   *Edited by Georgina Feldberg,
   Molly Ladd-Taylor, Alison Li, and
   Kathryn McPherson*

17 The Labrador Memoir of Dr
   Henry Paddon, 1912–1938
   Edited by Ronald Rompkey

18 J.B. Collip and the Development of
   Medical Research in Canada
   Extracts and Enterprise
   Alison Li

19 The Ontario Cancer Institute
   Successes and Reverses at
   Sherbourne Street
   E.A. McCulloch

20 Island Doctor
   John Mackieson and Medicine
   in Nineteenth-Century
   Prince Edward Island
   David A.E. Shephard

21 The Struggle to Serve
   A History of the Moncton
   Hospital, 1895 to 1953
   W.G. Godfrey

22 An Element of Hope
   Radium and the Response to
   Cancer in Canada, 1900–1940
   Charles Hayter

23 Labour in the Laboratory
   Medical Laboratory Workers in
   the Maritimes
   Peter L. Twohig

Labour in the Laboratory

Medical Laboratory Workers
in the Maritimes,
1900–1950

PETER L. TWOHIG

McGill-Queen's University Press
Montreal & Kingston · London · Ithaca

© McGill-Queen's University Press 2005
ISBN 0-7735-2861-X

Legal deposit second quarter 2005
Bibliothèque nationale du Québec

Printed in Canada on acid-free paper that is 100% ancient forest free (100% post-consumer recycled), processed chlorine free.

This book has been published with the help of a grant from Saint Mary's University.

McGill-Queen's University Press acknowledges the support of the Canada Council for the Arts for our publishing program. We also acknowledge the financial support of the Government of Canada through the Book Publishing Industry Development Program (BPIDP) for our publishing activities.

Twohig, Peter L.
  Labour in the laboratory: medical laboratory workers in the Maritimes, 1900–1950 / Peter L. Twohig.

(McGill-Queen's/Associated Medical Services (Hannah Institute) studies in the history of medicine, health, and society; 23)
  Includes bibliographical references and index.
  ISBN 0-7735-2861-X

  1. Medical technologists – Maritime Provinces – History – 20th century.
  2. Hospital laboratories – Maritime Provinces – History – 20th century.
  I. Title. II. Series

RB37.6.T96 2005   610.73'7'09715   C2004-906432-0

This book was typeset by Dynagram Inc. in 10/12 Sabon.

Contents

Figures and Tables   ix

Acknowledgments   xi

Preface   xv

Abbreviations   xvii

Illustrations   xviii

1 Introduction   3
2 Laboratory Work in the Maritimes: The Institutional Context   19
3 The Content of Laboratory Work   37
4 Not Just Bench Warmers: Labour in the Laboratory   58
5 Diffuse Roles and Multitasking   82
6 Recruitment, Mobility, and Wages   116

Conclusion   153

Notes   167

Bibliography   215

Index   235

Figures and Tables

FIGURES

5.1 Growth of the CSLT, 1937–1950 87
5.2 The First Two Hundred Persons Registered with the CSLT 97

TABLES

5.1 Geographic Origins of People Joining the CSLT by Province and Year, 1936–1945 100
6.1 Monthly Wages for Selected Canadian Hospital Workers, 1942 121

Acknowledgments

"The value of an Historian was discussed and agreed upon. No appointment was made at this time."[1]

Research and writing are selfish pursuits, but this book benefited from the help of many people and institutions. My partner in life, Beverly, never complained about the late evenings, the extended research trips, a two-year absence, or the missed family occasions. Our daughter, Sophie, has filled our house with delightful energy. Although I am certain she would prefer her books to this one, perhaps she will come to love Canadian history as much as I do. Bob and Gloria Younker fill our family life with love and I am grateful that they are my parents-in-law. My own father and mother have been good role models for the merits of hard work. I would like to dedicate this book to the memory of my mother, who died in June 2002.

Support of a different kind, though no less important, was provided by a number of funding agencies. The Social Sciences and Humanities Research Council, Associated Medical Services (through the Hannah Institute for the History of Medicine program), and Dalhousie University provided me with funding while I was a doctoral student. The Canadian Institutes of Health Research awarded me a postdoctoral fellowship that allowed me to complete this work, while two research grants from AMS, for another project, facilitated the discovery of some new material that I was able to incorporate here. I would also like to acknowledge the financial assistance provided by the Faculty of Graduate Studies and Research, Saint Mary's University.

The Canadian Society of Medical Laboratory Science (formerly the Canadian Society of Laboratory Technologists) was wonderfully supportive of this project. Kurt Davis and Lynn Zehr deserve particular mention. The national society made files available to me and provided office space while I was in Hamilton. I was also invited to present papers at the society's national congress in 1997 and also to a meeting of the Nova Scotia Society of Medical Laboratory Technologists. I appreciate very much the interest and support of the current generation of laboratory workers. In Nova Scotia, special acknowledgment is due to Dorothy Bidgood, who took an early interest in the book but died before I could complete it.

Michael Cross supervised the thesis on which this book is based. His dedication to his students, the stuff of urban legend at Dalhousie and beyond, proved accurate and I benefited from his criticism and support. He is a powerful role model, personally and professionally, and it was my privilege to be his student. I was also fortunate to have a wonderful group of other scholars to draw upon, including David Sutherland, Shirley Tillotson, John Farley, and T.J. (Jock) Murray. Each provided much-appreciated intellectual stimulation of differing varieties and amounts in their various areas of interest and, occasionally, mine. Colin Howell, of Saint Mary's University, reminds me through his work and friendship why I wanted to be a historian in the first place. Other colleagues at Saint Mary's have provided comfort and encouragement over the years. I would like to thank John G. Reid, Michael Vance, Madine Vanderplaat, Margaret Harry, and Terry Murphy for their support. Kathryn McPherson provided many constructive criticisms, while her own work has served as an inspiration. She has provided immeasurable assistance and encouragement during my effort to get a career established. The same is true of Jacalyn Duffin, one of the most generous scholars that I have met in the academic world. She is well deserving of her reputation as one of Canada's finest medical historians and one of the best mentors for students interested in the history of health. I owe all of these people significant intellectual debts and I know that I hope to emulate their generosity.

An earlier version of chapter 3 was published in *Acadiensis* and benefited from the comments of the anonymous reviewers of that manuscript and the former editor, Rusty Bittermann. Others offered feedback when I presented aspects of this work. The Dalhousie Medical History Society, active for more than a quarter of a century, provided a useful forum for testing some of my ideas. Papers were also presented at the annual meetings of the Canadian Historical Association and the Canadian Society for the History of Medicine. My colleagues in the Department of Family Medicine provided both intellectual stimulation and encouragement. Special thanks to Wayne Putnam, with whom I have shared a particularly productive working relationship.

The staff at Nova Scotia Archives and Records Management are simply wonderful. I think one would be hard pressed to find a more dedicated collection of individuals anywhere. John MacLeod, Barry Cahill, Lois Yorke, and Rosemary Barbour gave particularly helpful direction.

I also spent significant time in the Dalhousie University Archives, pestering the former archivist Charles Armour. He always treated me with courtesy and took an interest in my work. Staff at the Public Archives of New Brunswick, the Beaton Institute for Cape Breton Studies at the University College of Cape Breton, numerous university libraries, and others all helped me fill gaps in my research.

Aurèle Parisien of McGill-Queen's University Press took an early interest in the manuscript and provided much encouragement (and some prodding). Joan McGilvray provided key guidance, while Claire Gigantes was a supportive and expert copy editor. The two anonymous assessments provided by McGill-Queen's University Press helped me to refine the arguments. Responsibility for the remaining shortcomings is entirely my own.

Preface

The history of health care in the Maritimes is something of an orphan. While nationally great strides have been made in the field in recent years, including award-winning work by such scholars as Mary-Ellen Kelm,[1] few historians of the Atlantic region have turned their attention to the history of health despite a flourishing regional scholarship. There are relatively few studies of the hospitals, agencies, individuals, or professions in Atlantic Canada. There are several hospital histories,[2] a number of analyses of nursing education,[3] and other studies that explore aspects of health heritage during the twentieth century.[4] Often, health topics are dealt with only in passing in studies of other institutions, such as Peter Waite's history of Dalhousie or James Cameron's of St Francis Xavier.[5] We still lack good studies at the provincial level in this region, with the exception of Douglas Baldwin's studies of Prince Edward Island.[6] Little has been written about the many workers who comprise the modern health care complex, such as physio- or occupational therapists, x-ray technologists, dietitians, medical records clerks, or pharmacists, to say nothing of the porters, cleaners, and others essential to day-to-day patient care. This book attempts to put health care workers on the agenda through a case study of laboratory workers in the Maritimes.

I first became intrigued by the laboratory in Nova Scotia when I encountered a letter written by Dr Andrew Halliday, who was practising in Stewiacke. The young doctor wrote to the provincial medical officer, A.P. Reid, in 1901 to make inquiries about the position of laboratory director. Halliday wrote to Reid that he was interested because he enjoyed the "scientific side" of medicine and noted that he was "not strong enough for country practice."[7] I immediately framed several questions. Why was the laboratory established in the mid–1890s? What constituted the "scientific side" of medicine and how was it distinguished from the "clinical side"? Why did one have to be strong for country practice? Why was the laboratory a good refuge for a man like Halliday, who was not in good health and reportedly had a "feeble frame"?

This was good. But as I began reading about laboratories, it became clear that there were already many analyses of the physicians and research scientists who preached the gospel of scientific medicine and medicalized public health. Other voices, however, were muted. Increasingly, I became interested not in the origins of the laboratory or even the relationship between the laboratory and the medical profession but rather the staff who began to work in these settings and who have remained anonymous. In this circuitous and unexpected way, I came to the study of laboratory workers. This book is both an end and a beginning. It marks the fruition of a project originally undertaken for my PhD. Of course, occupational groups have particular histories with several unique aspects. But many of the issues that characterized laboratory work are broadly applicable to health care workers. I have therefore embarked on an explicitly comparative project of health care workers in the twentieth-century Maritimes.

Today, laboratory workers constitute the third largest health profession in Canada. This is probably surprising to many. The vast majority of them, roughly eighty percent, are women. I have uncovered evidence of their work in medical and nursing journals, annual reports of government departments and hospitals, and in the records of their national society. The result is this book, a case study of a single health care occupation in the first half of the twentieth century. Much of the case study is focused on Halifax, though I pay attention to other areas of the Maritime provinces as well. I believe that the issues raised herein hold important lessons for the writing of health care history throughout Canada.

Abbreviations

AGM	Minutes of the Canadian Society of Laboratory Technologists Annual General Meeting
AMA	American Medical Association
ASCP	American Society of Clinical Pathologists
BOC	Victoria General Hospital Board of Commissioners Minutes
CH	*Canadian Hospital*
CJMT	*Canadian Journal of Medical Technology*
CMA	Canadian Medical Association
CN	*Canadian Nurse*
CSLT	Canadian Society of Laboratory Technologists
CTA	Canadian Tuberculosis Association
DUAPO	Dalhousie University Archives President's Office Papers
JHA	Nova Scotia Journal of the House of Assembly
MHHC	Massachusetts–Halifax Health Commission
MMB	Minutes of the Medical Board, Victoria General Hospital
MMN	*Maritime Medical News*
NBARMH	New Brunswick Annual Report of the Chief Medical Officer to the Minister of Health
NSARM	Nova Scotia Archives and Records Management
NSMB	*Nova Scotia Medical Bulletin*
PANB	Public Archives of New Brunswick
PHAR	Nova Scotia Department of Public Health Annual Report
RT	Registered Technologist
VGHL	Victoria General Hospital Letterbooks

Laboratory, Hôtel-Dieu Hospital, Campbellton, NB. Nurse Irvine, Dr Felix Dumont, Sr Shannon. Courtesy of the Provincial Archives of New Brunswick, Collection: P23, Les Religieuses Hospitalières de Saint Joseph Photographs.

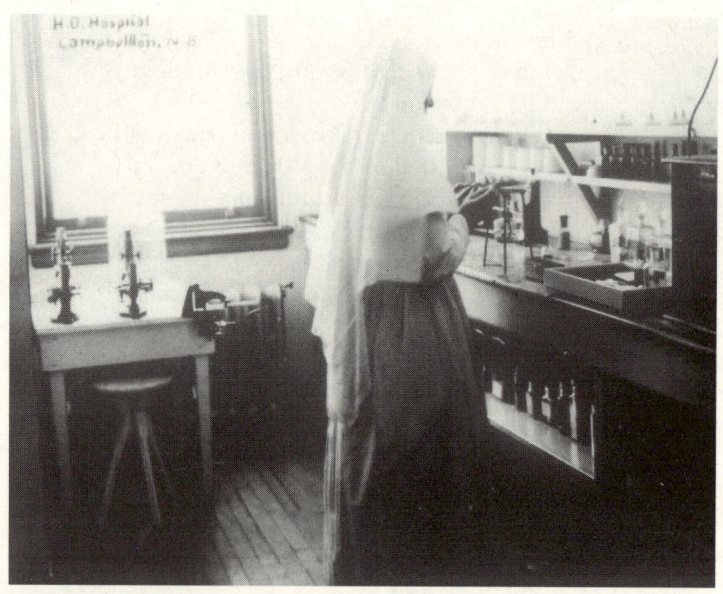

Laboratory, Hôtel-Dieu Hospital, Campbellton, NB. Courtesy of the Provincial Archives of New Brunswick, Collection: P23, Les Religieuses Hospitalières de Saint Joseph Photographs.

Laboratory workers on a Sunday afternoon picnic, ca. 1935. Margaret Low is second from left.

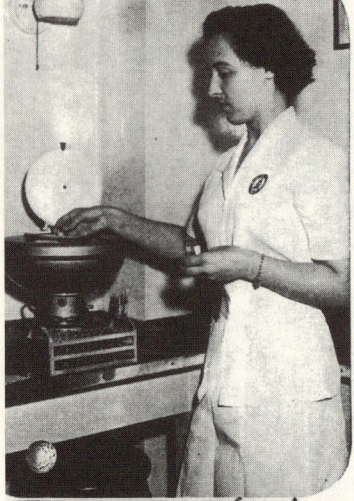

Testing Blood Serum

Each of the test tubes shown in the picture contain blood from a different patient.

The medical technologist is placing the tubes in a

CENTRIFUGE

which separates the cells from the liquid part (or serum) of the blood. In the SEROLOGY laboratory the student learns to examine blood serum for the effects of certain disease processes. The results of these tests assist the physician in the diagnosis and treatment of these diseases.

Cutting Tissues

In HISTOLOGY, tissues removed from the body in surgical operation are cut into very thin slices with a

MICROTOME

These are stained and mounted by the medical technologist and prepared for examination by the pathologist.

Description of laboratory tasks from a 1950s era career booklet.

Cover of CSLT Committee on Education career booklet, May 1952.

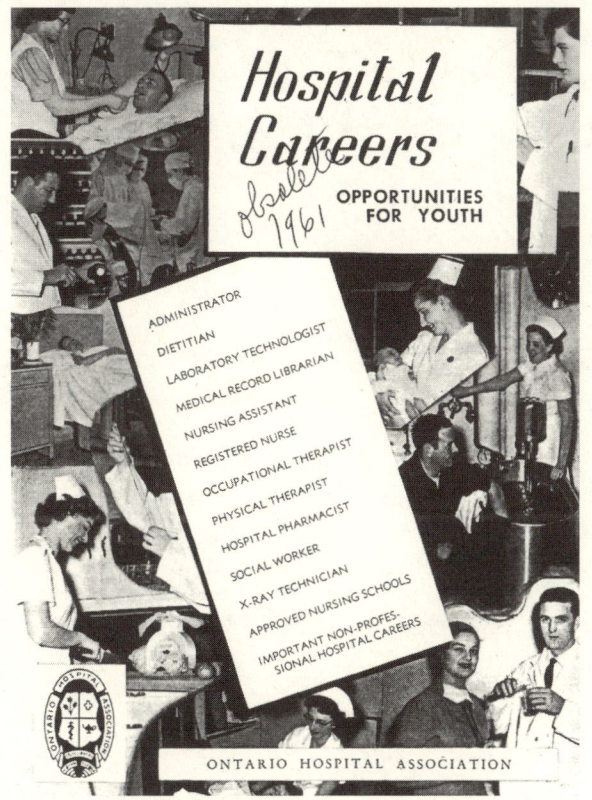

Ontario Hospital Association career booklet on hospital careers.

LABOUR IN THE LABORATORY

I

Introduction

How health professions work together, or fail to work together, is one of the fundamental questions facing the Canadian health care system in the current period of reform. In Canada, the introduction of health care providers such as nurse practitioners and licensed practical nurses and the "re-emergence" of midwifery have prompted questions about "appropriate" tasks, levels of education and training, and, of course, compensation.[1] Established professions are trying to reassert their jurisdiction over elements of health care delivery from which they have been alienated.[2] Even physicians are not immune to these pressures.[3] Creating, defending, and negotiating professional boundaries, in Canada and elsewhere, are contentious issues with deep historical roots.[4]

Interprofessional collaboration and multidisciplinary clinical teams are now commonplace in the health-care system. Some researchers cast this development as part of a larger transition from modernity to postmodernity.[5] The shifting environment of work practices and organizational structures within health care initiates a process of re-evaluation and reinvention for health service providers, wherein questions of professional identity arise. Central to this identity are the scope and limits of practice that differentiate one group of health care providers from another – the "boundaries" of the profession.

The social organization of health care workers is attracting the attention of policy planners, researchers, and government officials, all of whom share a penchant for multidisciplinary teamwork, integrated care models, or other such innovations. States are actively shaping health care delivery, rationalizing resources, facilitating the introduction of new kinds of providers, or changing the scope of practice of existing providers. Health care restructuring (one hesitates to use the word reform) prompts new relationships among different workers within the system.[6] Writing about new occupations, new tasks and responsibilities for health care workers, and issues of interprofessional conflict has become a virtual cottage industry among health service researchers.[7]

The prevailing climate of rapid change, encompassing shifting scientific knowledge, bureaucratic or administrative restructuring, technological innovation, and labour shortages, may act as a catalyst leading to negotiation or conflict among occupational groups. As evidence, we need only to think of overworked providers or underserviced rural communities. In both situations, the boundaries between providers become more permeable or disappear entirely. The busy rural family physician, eager to lighten the caseload, expands the role of other office staff. A rural community, anxious to ensure the presence of a primary-care provider, seeks a nurse practitioner.[8] "Upskilling," "multitasking," and "empowerment" are familiar terms, even clichés, within many health care settings. Across the health care delivery system, there are new opportunities for occupational groups to stake their claim to professionalism, for old ones to redraw boundaries or to reassert claims over areas of work they have lost.[9] The interdependence that characterizes health care work means that to understand what defines a profession, we must pay some attention to ongoing change in the context, content, and social organization of providing care. Any revision in the content of work for one group likely has consequences for others.

But this is hardly new. Work in health care has long been contested space and the boundaries between groups have often been subjected to competing, sometimes contradictory, pressures. The scope and practice of any one occupational group is frequently a matter for negotiation and renegotiation. Workers in health care also have myriad relationships and roles. Many are simultaneously members of a professional group and a labour union, or deliver some aspects of care while also serving as patient advocates, or work for wages while fulfilling managerial roles. Though there is much contemporary debate about occupational boundary maintenance within the health care system, there have been few historical examinations of the way professional boundaries were established, defended, and renegotiated.[10] As a result, occupational groups in health care have acquired an "aura of inevitable permanence," according to Larkin.[11] This book seeks to disrupt the idea of the immutable and permanent professional boundary, by looking at the emergence of one occupational group in a period of remarkable hospital expansion. The divisions between kinds of work in health care, the occupational boundaries, are of very recent origin. Indeed, what is striking about the experience of laboratory workers is that they illustrate how permeable professional boundaries actually were.

In Halifax, Nova Scotia, in 1919, one could pass by several hospital facilities walking along Morris Street in the city's south end, including the province's largest hospital, the Victoria General, and the children's hospital. On the northern side of the street was the stately Forrest Building, Dalhousie University's second home, while nearby, on the opposite side of Morris, stood the Pathological Institute. The "path lab" provided public health testing for the province and performed milk and water analyses for the City of Halifax and other municipalities. The laboratory also did clinical work for the Victoria General and other hospitals, for physicians in private practice, and for the Massachusetts-Halifax Health Commission, which was involved in many aspects of public health in the aftermath of the 1917 explosion in Halifax harbour.[12]

A few blocks away, on Cartaret Street, Margaret Low was hard at work. She was familiar with the complex of university and hospital buildings that were clustered along these few blocks. Low attended the medical college from 1900 to 1902, making her one of the pioneering women in Maritime medicine, alongside better-known names such as Florence Murray.[13] But in 1919, working from her home, Low was not providing medical care to the sick and infirm of her south-end Halifax neighbourhood. Instead, she was patiently cutting histological sections with a microtome, mounting slides, and taking them to the medical school, where students used them in their own studies, studies that Low did not complete.[14]

Margaret Low was also a pioneer in Halifax's Pathological Institute, though her contribution is largely unrecorded.[15] During her thirty years' service, she was a dynamic presence among laboratory workers. In 1927 the Department of Public Health sent Low to the Michigan Board of Health laboratories in Lansing, where she studied under Dr R.L. Kahn, one of the great laboratory scientists of the day. When the laboratory director, Dr D.J. MacKenzie, was absent for most of 1927–28 because of ill health, Margaret Low supervised the laboratory.[16] She was often responsible for training other women in the laboratory and was remembered by her co-workers as the unquestioned authority in the day-to-day work in the lab.

At the same time, Margaret Low exemplifies a great deal of what we know about women who worked and lived in Halifax. She was very well educated. She had attended the Halifax County Academy before entering university and pursued courses at the Maritime Business College afterward. Like many women in Canadian universities, she was underpaid. In Low's case, she earned an honorarium for her work, which she had performed since 1915.[17] University president A.S. MacKenzie noted that Dalhousie "could not put [compensation] on a really proper basis" but could "recognize that we appreciated her services."[18] While demonstrating in Dalhousie's laboratories, Low worked approximately four days a week for two hours a day in addition to preparing her slides. For her efforts she received one hundred dollars. This was considered sufficient until the late fall of 1919, when her brother died. Professor D. Fraser

Harris wrote that "as long as her brother was alive she had an income from him, but at his death that ceased. She is a[t] present doing so much for my department ... that it is only fair she should get something more than an honorarium."[19] So, in common with many women, Low's pay was not intended to provide a living wage.

Wage structures simply did not contemplate the independent woman, except for brief periods prior to marriage. Women were to eke out a living where they could find it. Laboratory work, however, offered women who needed and wanted work an alternative to tending to the needs of children, or family, or caring for patients (if one was fortunate enough to work exclusively in the laboratory). For educated women such as Margaret Low, this may very well have been a desirable occupation. They could acquire the necessary skills in fairly short order, often while being paid. If they had an undergraduate education in the sciences, laboratory work was a chance to pursue their interest, although in a highly constrained manner. At a time when employment opportunities for women in science were limited, the chance to use a degree that otherwise led to few career prospects, even for a short time, was desirable.

Margaret Low exemplifies the countless other workers who laboured anonymously in settings such as universities or hospitals. David and Rosemary Gagan have recently demonstrated how dependent the "science" of medicine and medical education was on "a growing hospital workforce of medical technicians and paramedical personnel," including laboratory workers.[20] Despite their significance in the emerging health care complex, we know very little about technical workers in hospitals and other settings. The experience of Margaret Low speaks to the invisibility of certain kinds of workers in our historical rememberings but they do exist in the documentary record. An examination of the Public Accounts of New Brunswick for the years 1919–20 to 1944–45 yields no fewer than 121 different names, including employees performing laboratory tests, support staff such as cleaners, clerical workers, and carpenters, and physicians working in the provincial Bureau of Laboratories in Saint John.[21] Still, such workers remain largely anonymous in most accounts of health care history.

The anonymity of laboratory workers has been shattered in recent years, as laboratories are once again in the public eye. Canada's tragic engagement with issues concerning tainted blood supplies and the devastating consequences for victims who contracted HIV or hepatitis C have thrust the laboratory into the spotlight.[22] In May and June 2000, when seven people died and 2,300 more fell ill because of E. coli in Walkerton, Ontario's water supply, questions about testing local water supplies took on national importance.[23] When cryptosporidium killed several and made many more sick in Battleford, Saskatchewan, in May

2001, the vulnerability of water supplies was confirmed.[24] Reports of suspect water supplies in other parts of Canada became common fare in the news. Water supplies were beginning to be regarded with caution, if not outright suspicion in many parts of Canada. More generally, questions surrounding emerging infectious diseases such as ebola or West Nile virus, drug-resistant strains of old nemeses such as tuberculosis, and, most recently, the anxiety over SARS (severe acute respiratory syndrome) have renewed the public profile of laboratories.[25] And while laboratories vary greatly in terms of their composition and objectives, they are all staffed by a variety of workers who toil anonymously, hidden from the headlines and historical accounts. The invisibility of laboratory workers is reinforced by the language of the hospital itself; samples are invariably sent "down" to the laboratory, regardless of its actual location.[26]

Stephen Barley and Julian Orr recently described technical workers as "the neglected workforce."[27] When power systems, water supplies, or other significant technologies are working, technicians are hidden from view. When problems occur, as they did in Walkerton or Battleford, these workers are thrust into the limelight. As Jeffrey Keefe and Denise Potosky noted, technical workers became visible "only when they made mistakes, departed from their assigned routines, or demonstrated incompetence."[28] Technical work is also poorly understood. In part, this is because of the broad nature of the workers' role. Some work with complex machines and systems that are rarely viewed by the public. Others attempt to ease our relationship with technologies that we use daily, such as computers, electronics, or automobiles. Technical workers are frequently encountered within health care. Sonographers, x-ray technologists, and laboratory workers occupy the interstices between physicians and patients. Technical workers also challenge traditional divisions between mental and manual work. Their work destabilizes easy dichotomies and makes the workers difficult to locate.[29]

Laboratory workers, currently called medical laboratory technologists, are such technicians. Overwhelmingly women, they are one of the largest groups of health care workers. There are more than fifteen thousand individuals registered with the Canadian Society of Medical Laboratory Science, working in hospitals and public or private laboratory settings. Medical laboratory workers work with human tissues and body fluids to diagnose or understand disease. They perform tests on this matter or prepare slides. Much of their work is highly automated, using the latest technology, while some work with specimens remains manual. Medical laboratory workers also have several areas of subspecialty, including histology (working with tissues), chemistry, hematology, microbiology, cytology, and blood bank.

Laboratories were an integral part of the rise of scientific medicine and the development of the public health movement that swept the western world in the wake of the great discoveries of bacteriology in the 1880s. However, we know very little about the internal workings of laboratories or their staff, particularly in the North American context. John Harley Warner surveyed the historiography of science and medicine during the 1970s and 1980s and concluded that many medical historians have displayed a reticence toward science.[30]

The lack of concern with science invariably leaves laboratories outside many investigations. Indeed, Andrew Cunningham and Percy Williams note that of the three sites of medical care, bedside, laboratory, and hospital, only the latter has received much historical attention.[31] Accompanying the transformation wrought by the new social history of medicine was a critique from within the medical community of reductionist medicine. Medical ethicists and clinicians jointly criticized the reification of medical technology for dehumanizing modern medicine by obscuring the doctor-patient relationship. Concurrent with this critique, resentment was building within the medical profession toward the research laboratory, both because of the latter's material support and its superordinate position vis-à-vis clinical judgment.[32]

Despite the problematic absence of laboratory medicine from the broader historiography of health, many studies have acknowledged that the laboratory was a key feature of medical education,[33] the location of medical authority,[34] and the site of medicine's most prominent discoveries, including the tubercle bacillus, diphtheria antitoxin, and others. A generation of scholars has pursued a program of research on the influence of laboratory teaching on medical education, of the laboratory-based germ theory on the history of epidemiology and public health, and of the increasing industrial role played by laboratories, especially (but not restricted to) the drug industry. But there have been few examinations of the laboratory itself, despite the institutional focus of much of the history of science and of medicine. In many hospital histories, the development of a laboratory is mentioned only in passing, if at all.[35]

But it is more than omission. When mentioned, laboratories tend to be portrayed in a specific way, usually through reference to their mysterious nature, equipment, and cost. In his classic examination of Chicago, Thomas Bonner mentions the laboratory when detailing the "meagre" equipment of County Hospital in the early years. The lack of laboratory (and other) facilities is cited as an indication that it was a premodern, poor-quality hospital.[36] The Hotel Dieu Hospital in Chatham, New Brunswick, reported to the health department in 1928 that the laboratory and other new services were an "immense expense" that contributed to the hospital's operating deficit.[37] William Warwick, New Brunswick's health minister, stated in his opening remarks to the 1936 annual report that the "highly scientific and exacting nature of the work carried on by the staff of the laboratory ... is little appreciated by the laity."[38] It is ironic, given their admirable effort to write history from the bottom up, that social historians have not seized on the laboratory in an effort to recover the narrative of these other workers and rethink many of the assumptions that endure about hospital and health care work. Perhaps the reason lies in the fact that social historians have viewed the laboratory as the preserve of elite scientists, thereby replicating the invisibility of the many technical workers; or perhaps historians of science have marginalized laboratory workers because they consider them to be unimportant.

Works that delve more deeply into the laboratory invariably focus on the elite research scientist. Even so recent and thorough an analysis as Gerald Geison's study of Pasteur fails to make anything but the briefest mention of workers other than Louis Pasteur. Geison describes Pasteur's laboratories as a "family affair," noting that the work environment emulated the family enterprises familiar in the French and Italian contexts. "Job security" characterized these labs, according to Geison, as members of the same family, extending into new generations, worked as "low-level technicians or custodial staff."[39] Despite his tease, these analytical paths are not pursued. In a similar fashion, Robert Bruce writes of anonymous "humbler workers" or "pyramids of pygmies," on whose shoulders the great scientists stood.[40] Bonner further obscures the history of the laboratory when he refers to the "bottle-crammed laboratory with its silent *men* of research."[41] The common view that the laboratory was a place of "scientific men, working in the scientific spirit," as the *New York Times* wrote when the Rockefeller Institute opened in 1903,[42] reveals as much about the historian as the history; it is rooted in assumptions of class and gender.

Other works, most notably Steven Shapin's *A Social History of Truth*, have endeavoured to make technicians visible, to transform them from the "ghostly inferred hosts of unnamed actors who shifted instruments about and exerted their muscular labour in making them yield phenomena."[43] Shapin has commented that technicians are triply invisible. First, they are virtually absent from the literature.[44] Second, the contribution of technicians to scientific pursuit is rarely preserved in the documentary record. Finally, stemming from the absence of technicians from the documentary record, it is plausible to infer that employers, including scientific institutes, hospitals, universities, and individual scientists, did not consider their work significant.[45] While we may accept Shapin's invisibility schema for seventeenth-century European science, it is far less tenable for the twentieth-century clinical or public health laboratory. Yet the same invisibility endures.

This book, then, seeks to focus some attention on this important segment of health care workers. In one sense, it fits within a broad international scholarship on professional groups. Whiggish accounts of professional formation, which dominated the sociology of the professions during the 1950s and 1960s, were roundly criticized in the 1970s for their uncritical acceptance of the professionals' own claims to exclusive knowledge and unique characteristics. A more fruitful path, identified by Everett Hughes, is to ask how occupations became professions.[46] With both the trait and process approaches, there was a search for essential attributes that proved problematic, as different groups followed unique paths, skipped steps, or failed to achieve others. There were variations according to setting. It seemed as if every group could claim professional status or none could.

A more productive analytical path was developed by critical sociologists during the 1970s. Rather than emphasizing trait or process approaches, critical theorists developed an analysis that placed control over the parameters of work at the centre. In this way, the conceptual link between "profession" and "occupational group" was disrupted. Not every occupational group was accorded professional status; professional groups were those with the ability to control their work and their membership. Concurrently, the concepts of "power" and "profession" became linked in most analyses developed since the 1970s. The ways in which power became formalized (say, through legislation) and institutionalized (in the form of regulatory colleges, for example) became the focus of neo-Weberian interpretations, including those of Eliot Freidson.[47]

While the critical examination of the professions that emerged in the 1970s did a good job of examining the ways in which class shaped professions, gender was left out of many accounts. Indeed, the idea of profession quickly became paired with that of the "semiprofession." The two key features of a "semiprofession," according to Amitai Etzioni, were that its members worked in a bureaucratic organization and that the members were primarily women.[48] In her examination of professional formation among physicians, midwives, nurses, and radiographers, Anne Witz concluded that the inattention to gender meant that "traditional and critical approaches to the professions continue to reproduce at the level of sociological knowledge professional men's own construction of their self-image."[49] To undo this tangle, Witz relies upon the "twin concepts" of patriarchy and professionalism.[50]

Witz's analysis of gender and radiography helps situate the experience of laboratory workers. Witz develops the idea of gendered strategies of demarcation, which "entail the encirclement of women within a related but distinct sphere of competence within a division of labour and, invariably, precisely because these occupations are gendered, involve their subordination to a male occupation."[51] Laboratory and x-ray technology are perfect illustrations of this process, although there are important differences between Witz's analysis of radiography in the

United Kingdom and my own analysis of laboratory work. Radiography in the UK was initially a mixed-gender occupation that, despite the efforts of male radiographers, was transformed into a "predominantly female occupation" by the mid-1920s. Witz describes the rapidity with which radiography was feminized and, through that, explores the failure of credentialist approaches to keep women out of the occupation. There is no similar process among laboratory workers; from the outset, laboratory work in Canada was overwhelmingly women's work.

To understand laboratory work, it is also necessary to think not only of the occupational boundaries and demarcationary strategies described by Witz but also of the permeable nature of those boundaries. Witz argues that "the boundaries between occupations are constructed as gendered boundaries,"[52] and, while certainly true, it is equally true that the portals between occupational groups were also gendered. This is to say that the expanding services of the modern hospital depended upon the talents of multitasking women, a pattern that prevailed in many settings for much of the twentieth century. In this way, I want to develop further two of the analytical points raised through Witz's study of radiography. She explains radiography's feminization through three interrelated explanations.[53] Firstly, Witz argues that women had gained some experience with x-ray equipment through their capacity as nurses. Secondly, x-ray training was hospital based and women already occupied spaces within the hospitals, replete with their own training schools. Through the 1920s and 1930s, the staffing needs of Canadian hospitals expanded enormously, while most hospitals had only a limited fiscal base. For many hospitals, nurses were not only an appropriate labour supply for the new diagnostic services but also a fiscally sound one.

The hospital division of labour and the gendered spaces it created provided some new opportunities for women, though firmly within a medical hierarchy. As Witz demonstrates, male physicians increasingly articulated the medical division of labour in gendered and patriarchal terms that favoured the employment of women. Here, we can turn to the work of Tracey Adams who has examined such patterns through the case of dentistry in Ontario. Importantly, this analytical path inexorably leads one to consider the ways in which women were *included* in the world of professional work, albeit streamed into auxiliary work. The employment of dental assistants helped the professional project of dentistry by freeing men from the more manual/technical and lower-status aspects of their work. Women assistants were also inexpensive and could fulfill clerical duties. Finally, women assistants were considered to be subordinate and tolerant of male authority. Indeed, dentists were warned against training male dentists who, it was thought, might reject the authority of dentists and enter into competitive practice.[54]

In other ways, the emergence of the laboratory labour force parallels that of radiographers.[55] Radiographers (x-ray technicians in the North America in the early twentieth century) emerged alongside the development of radiology as a medical specialty. As Gerald Larkin has ably demonstrated, the intellectual work of diagnosis from x-rays became vested in the physician, while the technical work of producing x-rays became the responsibility of the radiographer.[56] Witz identifies this difference as the key element of radiologists' demarcationary strategy. Here there are strong parallels between the laboratory and the x-ray room. The manual work of the diagnostic services was abandoned by medical men, defined as "scientific" but manual, and deemed to be appropriate for women. It is clear that technological innovations and shifts in the labour process of health care created new spaces for women. It is equally clear that this transformation affected women differently from men. Male physicians were firmly ensconced at the top of the health care hierarchy, their claims to skills confirmed and enhanced by the presence of a variety of auxiliary workers, including those who laboured in the laboratory. Increasingly, the craft of health care was divided among various groups, with physicians laying claim to the intellectual work of diagnosis and leaving the more routine or technical tasks to others.

Professionalization is a useful framework through which to understand laboratory work but it is not the only one. There is also the large and lively scholarship on women and technology, which includes Elizabeth Baker's classic *Technology and Women's Work* and the work of Ruth Schwartz Cowan.[57] The result of this work is varied, ranging from liberal concerns about access to such fields to radical assessments of the application of technology. In part, this literature drew upon a feminist critique of professional culture (including academic culture) that emerged in the 1960s and a critique of science and its authority by social scientists. Technology studies quickly became an interdisciplinary field as scholars came together, fused by their common criticism, or at least the shared object of their criticism. In laboratory work science and technology are in close proximity. The knowledge developed by science is given its practical application. Of course, science and the "scientist" are decidedly masculine images. Technology is more problematic, conjuring up images of masculinity or femininity depending upon context. Perhaps this ambiguity explains why laboratory work avoided sex typing to a certain extent, despite the overwhelmingly female composition of the labour force in North America.

There is, however, abundant evidence of sex segregation in the health care labour market. In her examination of technological change in printing, clothing manufacture, the filling of mail orders, and x-ray work, Cynthia Cockburn examines how gender and class relations interact to exclude women from skilled jobs, with the result that they are shunted off to other occupations. In the pages that follow, there is the example of Margaret Chase, a physician who goes to work in a Halifax laboratory the day after her graduation, and the consideration of two female physicians for a technician's job in Prince Edward Island. There are other examples of women doctors who were directed into institutional work, rather than private practice. Many other women were directed to laboratory work as a "good career." There are examples herein of women who were encouraged to pursue further training or to take additional courses in the medical sciences to prepare them for work in the laboratory. There are no examples of such support or encouragement to prepare for, say, a career in medicine or a career in research. The exclusion of women from these occupational domains created the conditions that underpinned their entry into laboratory work.

Any understanding of the nature of health care work is incomplete without paying some attention to the material conditions that shaped the development of hospitals and public health, the key employers of laboratory workers. That is, the interplay of gender and professional formation did not operate apart from other factors such as the ways in which hospitals were funded, the level of that funding, the drive toward hospital standardization, and a range of other factors. In Canada, health care before 1950 was largely the responsibility of municipalities, with varying amounts of provincial support. There is therefore a profound need to situate interpretations of professional and occupational groups within the historical and structural parameters of the development of health services and, through this, to ground them within the larger processes of welfare capitalism. Neo-Marxists such as Magali Sarfarti Larson argue persuasively that it is not enough to view professional formation simply as a matter of occupational closure.[58] We are better served exploring such groups for what they reveal about political, social, economic, and gender relations.

This book is a case study of laboratory work and workers in the Maritimes, but one that raises questions that are significant both in the rest of Canada and in other settings. It is important to acknowledge at the outset that, while I consider this to be a regional study, it is not *merely* a regional study. It is certainly true that developments in the laboratory were influenced by trends elsewhere, but, to cite Wendy Mitchinson, "to see [such developments] only within an international context is, I think, a rather provincial and limited view."[59] As Mitchinson argues, there is a need for Canadian studies of aspects of health care because place matters. I would extend this argument to regional or local levels of analysis. My approach in this book is similar to that of other scholars of science and medicine, who carefully situate their studies in specific settings. Some of the most widely cited analyses of hospitals, laboratories, and medical technology ground their analyses in the particulars of a community setting. As illustrations, we need to think only of David Rosner or Morris Vogel's classic studies of hospitals in Brooklyn and Boston respectively, Bruno Latour and Steve Woolgar, who studied the Salk Institute at La Jolla, California, or Joel Howell, whose *Technology in the Hospital* drew largely upon evidence from Philadelphia and New York.[60]

Such studies contribute to a robust analysis of science and medicine. There is, as Ian McKay recently wrote in *Acadiensis*, an "empirically strong case to be made for a regional framework of analysis" for many topics and surely health care is one of them.[61] Local government has always played an important role, for example by directly funding infrastructure such as hospitals in different ways and at different levels during the first half of the twentieth century.[62] Some municipalities built infrastructure for water supplies and sewage disposal or enacted bylaws governing such areas as refuse disposal or sanitary inspection of restaurants, abattoirs, or dairies. Provincial governments in the Maritimes and elsewhere in Canada funded "provincial" institutions such as large general hospitals, sanatoria, mental hospitals, and laboratories, in addition to a range of public health activities.

Paying attention to the local experience is critical in other respects as well. One of the key points that I wish to make is that the health care system in Canada was, and remains, idiosyncratic. Individual institutions and provincial health departments, even if broadly comparable, showed incredible variation in how they were organized, staffed, and administered. Training programs for health care professionals and licensing requirements also show considerable variation. I have therefore attempted to situate key aspects of the experience of laboratory workers in the Maritimes within a national framework. There is no doubt a need to understand the development of health care workers in many different settings. As the first study of medical laboratory workers, my book marks a beginning and I hope that it helps to reshape our historical understanding of the history of health care workers in Canada.

This study is primarily interested in the laboratory as a place of work, not as a place of science or even of medicine. I therefore pay very little attention to the international development of the various diagnostic tests, vaccines, or serums. There are many accounts that cover this territory.[63] Clearly, then, this is not a patient-centred study. What the various tests meant for patient outcomes or for the health of the province is not examined here. Much of the evidence I present is drawn from Halifax, though other Maritime laboratories are included. In attempting to situate the experience of laboratory workers, I emphasize their relationships with other workers in the institutional context.[64] It was certainly not my intention to write the history of a particular laboratory or hospital, but it has been necessary to provide some detail on institutional development. I therefore describe the development of the local laboratories in the next two chapters, deferring discussion of the women who laboured in them.

This is a necessary evil, given our limited knowledge of the establishment and functioning of Canadian laboratories. In the absence of a well-known and robust historiography, replete with standard reference works that can be plucked off the shelf, Canadian historians of health care often have to tell multiple stories in their studies: we still often work without a historiographical net.[65] At some level, all historians of health care have to reconcile their desire to analyse a particular historical experience in a particular setting with the demand to make a contribution to our understanding of Canada and of the development of health care elsewhere.[66]

Laboratories developed in Canada as part of the public health reform effort. By 1919, public health in Canada had been transformed, according to Paul Bator, "from the status of a periodic preoccupation of a few doctors and lay volunteers into a permanent occupation for experts who daily monitored the health of communities."[67] From about 1880 to 1920, municipal and provincial governments established public health boards and departments across Canada.[68] Ontario passed the first legislation in 1883, followed by Quebec in 1886, New Brunswick in 1887, and Nova Scotia, Manitoba, and British Columbia in 1893.[69] New Brunswick appointed the first minister of Public Health in Canada in 1918, and the federal government created its own Department of Health in 1919, though it had a long-standing interest in some aspects of health, notably the health of First Nations.[70] Taken together, these developments put in place the basic administrative framework for public health.

The involvement of federal, provincial, and municipal governments, undertaking a variety of initiatives and developing in different ways, made public health terribly complex. Many individuals and occupations were active in defining public health issues and designing solutions. Developments in chemistry and physics made some public health efforts viable, while engineers refined sewage systems and water supplies. In the twentieth century, nurses assumed an increasingly prominent role in the public health effort. Large organizations, like the Canadian Tuberculosis Association and community service clubs such as the Kiwanis or Women's Institutes, took on specific challenges within public health. Chapter 2 explores the complex relations among different groups that were found in early Maritime laboratories. Descriptions are offered of the particular development of the Pathological Institute in Halifax and Saint John's Bureau of Laboratories. Initially, both were small facilities with modest equipment, hardly exclusive bastions of "scientific medicine" divorced from clinical care. Indeed, both facilities cultivated strong working relationships with community physicians and hospitals around their respective provinces. During the formative years between 1910 and 1930, laboratory development in Saint John and Halifax required the broad participation of a number of constituencies. In this way, the history of laboratories paralleled other developments in health care that were mounted by provincial governments in the region only after commitments were secured from other parties, including municipal or federal governments, philanthropic foundations such as the Rockefellers or voluntary agencies such as the Red Cross.

Health care innovations were driven by the horror of disease. Typhoid outbreaks initiated calls for better water and sewage disposal, smallpox prompted vaccination efforts, diphtheria spawned antitoxin (and later toxoid) campaigns and tuberculosis shaped a variety of medical and social responses. The burden of disease, discussed in the second chapter, defined both public health work and the work of the laboratory. The laboratory aided diagnosis of a number of diseases by conducting tests and reporting findings to local physicians or public health authorities. One of the largest areas of activity in the laboratory during the formative years was the effort to combat venereal disease. More importantly, the "VD problem" captured the federal government's interest and Ottawa committed resources that permitted the expansion of provincial laboratories.

By the 1920s, the expanding workload brought many different workers into the laboratory, as discussed in chapter 4. Nurses were among the earliest to assume duties in the laboratory in addition to caring for patients. Indeed, many nurses filled a variety of roles in their hospitals, working in a number of services and departments, including laboratories, x-ray departments, dietetics, medical records, or the pharmacy. The size, location, and nature of the hospital were key determinants of whether a nurse would have to fulfill multiple roles. The importance of setting is made clear by an exploration of Dalhousie University's medical science laboratories. Renewed in the 1920s through donations from American philanthropist, these laboratories were largely staffed by young men and serve as an important reminder that jobs were gendered in equally important ways for men and women. In contrast, the staff of the large provincial laboratories in Halifax and Saint John were overwhelmingly single women.

Early laboratory workers had diffuse roles, readily moving between tasks such as typing reports, setting up or cleaning apparatus, and performing tests. Like the nurses described in chapter 4, they also worked in a variety of departments in the modernizing hospital. Chapter 5 explores the many paths that led to the laboratory and the ways in which the Canadian Society of Laboratory Technologists (CSLT), the professional body founded in late 1936, accommodated individuals with varying educational qualifications and work experience. One of the chief efforts of the early CSLT was to create a national curriculum and establish training programs, two of the hallmarks of professionalism. Debates about what education, or how much, was appropriate to prepare for a career in the laboratory reveals much about the emerging professional identity.

Finally, chapter 6 examines several challenges that confronted laboratory work and its emerging identity in the 1940s and 1950s. Most critical in the Maritimes were issues of salaries, mobility, and retention. The lure of other locales was particularly profound and many workers left for other parts of Canada or for the United States. This issue was complicated by the difficulty of recruiting people to the profession and a rising demand for workers during World War II and the immediate postwar period. Finally, some consideration is given to the professional portrait of laboratory workers that emerged in the postwar period.

NOTE ON SOURCES

The question of sources is often a thorny one for historians working in the twentieth century. There are journals that offer published accounts of the national scene, such as the *Canadian Journal of Medical Technology* or *Canadian Nurse*. There are also the annual reports of departments of health, which took an early and active interest in not only public health laboratories but smaller clinical facilities as well. These sources, however, are not entirely satisfactory for understanding the workers at the bench. Published accounts of individual laboratory workers are virtually nonexistent, and they lack the prominent leaders of other professions, such as nursing. For this study, the richest sources are held in the national office of the Canadian Society of Medical Laboratory Science, where they have been fortunately preserved for posterity. The executive of that society (formerly the CSLT) graciously granted access to the membership files, the minutes, and other administrative records. For those workers who attended university, student records were available. Finally, eleven of the earliest workers in Halifax and elsewhere granted oral interviews that helped to frame my understanding of laboratory work.

The approach to oral interviews was straightforward. Potential participants were identified through various means and contacted by a letter that explained the project and asked whether they would agree to be interviewed. Those who were willing to recount their experience were interviewed and audiotaped. Interviews were guided by a series of questions and the knowledge already gained of their work experience. Often, questions could be posed about co-workers or developments because of the information found in primary documents. This often aided the discussion. Individuals were given an opportunity to recall events that they imbued with significance. For many, the years before World War II were dark days, when their work was rudimentary and their careers were in the early stages. For others, their tenure in the laboratory was a small and perhaps insignificant part of their life. Such is the nature of the work. Many stayed in the laboratory for only a brief period before moving on to other careers or to marriage and child rearing.

There are limitations to this approach to be sure. It is neither systematic nor representative. I did not actively seek women from a variety of backgrounds for the interviews and instead used a convenience sample. There is a strong bias toward those who were university educated because they were more easily traced through university alumni records. But using an interview guide and applying insights learned through other kinds of evidence enabled the interviews to be contextualized. More remains to be done in capturing the stories of these women. But the modest beginning contained herein enriches the documentary record and offers insights otherwise not preserved. An employment application may tell the historian about a person's abilities or education but reveals little about what prompted that person to pursue a particular field. Oral evidence does more than offer illustrations to support or confirm other findings: it offers new insights. While the information in these records is evocative, access arrangements with various organizations required confidentiality. Thus, in the chapters that follow, I have used pseudonyms to protect confidential information. When information was derived exclusively from publicly available sources, as in the case of Margaret Low, I have maintained the real name.

This study attempts to apply the insights of social history to this important group of health care workers. That they should be overlooked, despite being the third largest health profession in the country, is not entirely surprising. The great discoveries of the laboratory are usually portrayed in discovery myths as the work of a few scientists. The people who make the work possible, in common with those who work a rockface in a mine or a loom in a factory, are invisible. The faces the public remembers, the figures who continue to dominate our historical memory, are those who accumulate wealth or lend their names to buildings, institutions, enterprises, or discoveries. It is the physician who cures, the scientist who advances knowledge, or the government that ensures clean water. That others strive behind the scenes is, more often than not, obscured. Stating the obvious, the cumulative effect of these omissions bears heavily upon the history of women. This study will attempt to recover the contribution of this hitherto unexplored group, women laboratory workers.

2
Laboratory Work in the Maritimes: The Institutional Context

To understand laboratory work, it is necessary to understand the development of the institutional structures. As suggested in the introduction, there have been few studies of the laboratory in Canada. It is therefore necessary to provide some description of how and why laboratories were created in the Maritimes. This chapter explores the development of institutions, leaving the discussion of the workers to subsequent chapters. Institutions provide the context for work, creating relationships among staff members, among services, or between settings. Laboratories in the Maritimes were not among the earliest and were not innovators. Instead, they were created in a national and international climate that saw laboratories established throughout North America and western Europe in the closing decades of the nineteenth century and the early twentieth century. The leading laboratories in the Maritimes were the Pathological Institute, located in Halifax, and the Bureau of Laboratories in Saint John, New Brunswick. These laboratories were utterly ordinary, but they became important centres for public health, for aiding physicians' clinical work, and, in Halifax particularly, for medical education.

Laboratories were an overt and visible manifestation of the broad development of public health in Canada.[1] Ontario established its public health laboratory in 1890, eight years after creating a provincial health board.[2] In Montreal, Dr Louis Laberge, the medical officer from 1885 to 1913, laid the foundations for a laboratory service.[3] In these and other laboratories, scientific work in bacteriology, microbiology, immunology, and biochemistry was given practical application. Large municipalities and provinces were quick to recognize the utility of laboratory tests. Serious and common contagious diseases such as diphtheria and tuberculosis could be better diagnosed with the assistance of laboratory tests.[4] An early annual report from Nova Scotia noted that the laboratory would "be of special service to the [medical] profession and the public, for an early diagnosis in cases of diphtheria and tubercle, when the signs are indefinite ... Few medical men have the apparatus, even if they have the skill, to make such examination in doubtful cases. In very many cases doubt has been dispelled, and in others a timely forewarning given which will save, or in any case, prolong life."[5] By the early twentieth century, the laboratory could aid the diagnosis of many diseases and offer assessments of milk and water purity.[6]

In Nova Scotia, the origins of laboratory service are murky. In a brief article, D.J. MacKenzie suggests no less than three dates for the origin of laboratories, all from the mid-1890s.[7] Imbued with the spirit of public health reform, particularly with issues of sewage disposal, potable water, safe milk, and unadulterated food, a legislative committee recommended that a laboratory be outfitted and placed under the direction of a bacteriologist. The laboratory would supply diphtheria antitoxin, smallpox vaccine, examine and report on sputum samples for tuberculosis, and conduct bacteriological examinations of water, milk, and food in suspected cases of contamination. The members of the house unanimously accepted the need for a laboratory.[8]

Despite the enthusiasm of the legislature, only a rudimentary laboratory was established. A committee of "medical men" recommended Dr W.H. Hattie for the position of provincial bacteriologist at an annual salary of three hundred dollars. The committee allocated a further one hundred dollars for equipment.[9] From the outset, there was a gap between the government's stated commitment to public health and the resources it was willing to provide. Such gaps were characteristic of many of the early health initiatives in the Maritimes. While publicly committed to laboratory expansion, hospitals and public health authorities sought ways to economize. As we shall see in subsequent chapters, this had profound implications for laboratory workers and other hospital staff.

Hattie served as director for six years, until a new laboratory opened in August 1901. By then, the peripatetic service even of someone as accomplished as Hattie was not sufficient to meet the growing needs of the facility. The new laboratory was a modest facility located in the Halifax Medical College. The laboratory was administered in connection with the provincial medical board and placed under the direction of Dr Andrew Halliday.[10] Halliday was a native of Hutton, Scotland, who came to Nova Scotia in 1892. He was a graduate of Glasgow University and practised in Stewiacke. Halliday enjoyed scientific work and he maintained a small laboratory in addition to his rural medical practice.[11] He taught one session at the Halifax Medical College and that was enough to convince him to devote his time and career to laboratory medicine. Halliday spent a year in Durham, England, pursuing postgraduate studies in bacteriology, pathology, and other public health courses. It was while he was in Durham that the young doctor first approached Dr A.P. Reid, the secretary of Nova Scotia's board of health, about the position of laboratory director.[12] By 1901 Reid believed that Halliday was "just the man N.S. wants," adding that "Dr Hattie has too much to do."[13] Upon his return to the province, Halliday was appointed associate professor of pathology at the Halifax Medical College and director of the laboratory.[14]

Halliday was to serve both individual clinicians and the public health needs of the province. While the public health role was clearly emphasized, the government made explicit the laboratory's role in assisting physicians throughout the province in their clinical work. He would also act in an advisory capacity on matters of public health, offering counsel to the provincial and municipal governments as they attempted to establish more robust public health programs. Halliday's multiple roles blurred the divide between clinical work and laboratory-based medicine, which had the salubrious side effect of easing any remaining tensions between scientific and bedside medicine. Initially, and not insignificantly, laboratory analyses were portrayed as an adjunct to the clinical acumen of the attending physicians. By the early twentieth century many physicians acknowledged that laboratory testing, and other technologies such as the x-ray, could reveal health problems overlooked by the clinical exam. Yet they were also mindful that diagnostic technologies could alienate a fundamental aspect of clinical work from the practising physician. This was not a simple matter of fearing for one's livelihood. Rather, it represented a fundamental difference in philosophy. While advocates of laboratory medicine believed that laboratory tests or other diagnostic technologies provided definitive answers, clinicians emphasized the continuing importance of clinical judgment.

Ill health forced Halliday to resign as laboratory director in 1902 and he died of tuberculosis on 19 March 1903, at the young age of thirty-six.[15] Despite his brief tenure, Halliday brought a level of expertise and commitment to the laboratory that had a lasting impact on the medical community in the province. The work grew steadily under his tutelage, and the number of specimens and samples sent in mounted. While eulogizing the esteemed doctor in the *Maritime Medical News*, it was remarked that the "necessity for a Provincial Bacteriologist is now established beyond question."[16] Throughout Nova Scotia, physicians were growing convinced of the utility of the laboratory. This was particularly true of those who, like Halliday, showed interest in the science of medicine. The point of departure for many in the tension between the art and science of medicine was whether or not clinical skills would prevail over diagnostic technologies. For the vast majority of practitioners they would; laboratory tests or x-rays were only aids to clinical abilities. By positioning themselves as the ones who ordered and interpreted the meaning of tests for patients, physicians ensured that their judgment would remain supreme.

THE PATHOLOGICAL INSTITUTE, HALIFAX

Laboratory testing was growing in importance, but the early history of the laboratory in Nova Scotia was characterized both by shifting locations and changes in personnel. Dr L.M. Murray replaced Halliday, who had been his mentor. Murray augmented his education with further courses at McGill University and American centres such as Washington and New York.[17] Each year Murray reported that the laboratory was performing more, and more complex, tests. In 1904, for example, new equipment was added to facilitate pathological investigation of tumours and tissues. By 1908–09 Murray had requested additional equipment and staff to keep pace with the expanding workload.[18] In 1910 the laboratory moved to new quarters in the Technical College of Nova Scotia. This was only a slight improvement because the entire laboratory still consisted of just one room, but it was more satisfactory than the accommodations at the Halifax Medical School.[19]

A better facility was needed, however. In early 1911 the board of commissioners of the Victoria General Hospital began discussing the possibility of building a new pathology laboratory and securing the latest equipment. The Liberal government of Premier George Murray even placed the proposed new laboratory in the budget estimates for the next fiscal year.[20] From the outset the new facility was to meet the clinical needs of the hospital, the pedagogical needs of the university, and the public health demands of the city, province, and, periodically, even the federal government. The committee selected a site on Morris Street, close to the hospitals and to Dalhousie University.[21] The brick two-storey building cost approximately $23,000 and officially opened on 1 March 1914.[22] W.W. Kenney, the hospital superintendent, claimed that "there are few better laboratories in Canada."[23] A new building with new equipment in a rapidly changing context of medical science probably did much to bolster Kenney's claim. In appearance, at least, Nova Scotia had turned the corner and was ripe for a robust program of health.

In advance of the laboratory's opening, a search began for a "bona fide director."[24] Dr M.A. Lindsay of Birmingham, England, who had earlier made inquiries about securing an appointment in Halifax as pathologist, was the leading candidate. Negotiations were conducted throughout the month of May 1911 by means of cablegram, with Lindsay eventually agreeing to an annual salary of two thousand dollars in addition to consultation fees.[25] This too was an important consideration. The laboratory director would not depend solely on the government for his income, and this preserved his professional identity and independence. He would remain a member of the community of physicians earning at least part of his living through fee-for-service. Like his predecessors, Lindsay would be responsible for the laboratory work for the province, City of Halifax, Victoria General Hospital, and some university work, including teaching. For its part, the hospital reserved the right to expand the work beyond that of the hospital. Kenney reiterated that the laboratory was to meet "not only the requirements of the hospital, but also of the Province and the City, and possibly Nova Scotia, and possibl[y] the College."[26] Lindsay was to have had a broad role in bringing laboratory science to the medical profession of Nova Scotia, the medical students, and public health officials. His task was cut short, however, when he died in the *Empress of Ireland* disaster in 1914.[27]

Following Lindsay's death, the laboratory itself became a key means to attract a replacement. Writing to the superintendent of Johns Hopkins Hospital in Baltimore, W.W. Kenney suggested that the new lab would be entirely satisfactory to a qualified director.[28] In the interim, Dr L.M. Murray once again assumed those duties.[29] Many candidates were again considered before the position was offered to Dr A.G. Nicholls of Montreal.[30] Nicholls had been born in England, although he was raised in Montreal. He graduated from McGill's Faculty of Medicine in 1894. Postgraduate study took him to Europe and back to McGill, where he eventually earned a Doctor of Science degree. Before coming to Halifax, Nicholls had been a professor of pathology and bacteriology at McGill, and the assistant pathologist at the Royal Victoria Hospital.[31] He enjoyed some standing in Canada as "an acknowledged authority in pathology and bacteriology"[32] and was appointed at a salary of twenty-five hundred dollars, to which Dalhousie University made a five-hundred-dollar contribution. Like Lindsay before him, Nicholls was to be responsible for a broad range of activity and was permitted to charge fees for consultations.[33] The investment was a sound one. Nicholls remained in the position for more than a decade, and presided over the expansion of the laboratory service in Nova Scotia. Although there was further physical displacement in the 1920s due to the expansion of the Morris Street building, the diagnostic laboratory had finally found a permanent location and a long-serving director.

The establishment of the laboratory service is best viewed as part of what Ian McKay called the "progressive zeal and wave of reform"[34] that characterized the 1910s. Interest in health matters broadened greatly and the pace of reform quickened. The Liberals included such measures as workers' compensation, a contributory old-age pension plan, and factory regulations in their electoral platform during the campaign of 1911.[35] At the same time, a period of transition within medical education in Halifax culminated with the reintegration of the Halifax Medical College into Dalhousie University.[36] There were campaigns against tuberculosis, such as the one conducted by the Tri-County Anti-Tuberculosis League in eastern Nova Scotia,[37] and other manifestations of the reform impulse such as the Jost Mission,[38] the antidrink effort, and the suffrage campaign.[39]

There were health initiatives throughout the Maritime provinces. New Brunswick established its Department of Health in 1918 and there were limited developments as well on Prince Edward Island.[40] In Nova Scotia, the province attempted to introduce public health nurses, though the experiment met with only limited success.[41] Dalhousie embarked on a brief period of training public health nurses in 1920.[42] The program, like many of the health initiatives, did not endure. The 1910s brought growth to the Maritimes, though there were significant problems in the primary sectors. The economic future nevertheless looked bright. In sharp contrast, the 1920s were a time of regional economic crisis. Employment in the fisheries declined from a peak of 17,583 in 1919 to only 12,395 by 1923. Within two years, between 1919 and 1921, employment in the manufacturing and mining sectors dropped from more than 46,000 to under 28,000, and by 1925 the value of production was less than half of its 1919 total.[43] In this context it is not surprising that the reform impulse of the 1910s abated.

GROWTH AND REORGANIZATION AT THE PATHOLOGICAL INSTITUTE

The expansion of the laboratory through the 1910s illustrates how fragile "progressive" reforms were in Nova Scotia. Broad participation and money were required to modernize many aspects of health care in the Maritimes.[44] On a practical level this meant that the provincial government, individual municipalities, and Dalhousie University were all interested in the operation of the Pathological Institute. The laboratory also had to fulfill a number of objectives, including providing public health tests, aiding clinical diagnosis, addressing social issues, and providing opportunities for medical education. The variety of interest groups and competing claims on the work of the laboratory brought administrative difficulties. As early as 1909, the Victoria General appointed a special committee to manage the laboratory.[45] During late

1910 and early 1911, as the medical college was struggling to define a place for itself, Dalhousie sent delegations to discuss a new relationship with the laboratory. Nothing was settled since it was a period of both laboratory expansion and medical school reform, but the hospital commission agreed to give some consideration to the pathologist's teaching role and whether the new laboratory might be opened to the medical students.[46]

In 1911 the board of commissioners requested that the medical board further delineate the role of the pathologist. The medical board made eight recommendations, notably that the appointee's services "shall be at the disposal of the Commissioners and he shall not undertake any other work without the consent of the Commissioners."[47] When Dr Lindsay was appointed to the facility in 1911, he was permitted to consult with other physicians and charge fees over and above his two-thousand-dollar annual salary.[48] In allowing the laboratory director consultation privileges, the board ensured that the laboratory and its director would become an accepted part of the medical community, building on the earlier decision to provide clinical services to physicians in private practice throughout the province.

In August 1911 yet another Dalhousie delegation met with hospital officials and agreed to contribute three hundred dollars per year to help defray the construction and operating costs of the laboratory, in exchange for access to the facility for teaching purposes.[49] For its part, the hospital agreed to provide only the physical environment and consistently refused to outfit the students' laboratory with microscopes and other moveable equipment.[50] In correspondence with Dalhousie president A.S. MacKenzie, Kenney suggested the hospital could not justify "undertaking such an uncertain, and more or less continuous expenditure, as would necessarily be involved in providing and maintaining such equipment." The agreement also specified that the laboratory room was to be used only for teaching and only during "regular teaching hours" and that permission from the board of commissioners was required for anything beyond that.[51] Occasionally, Dalhousie placed even more demands on the laboratory. In 1916 MacKenzie asked that the hospital waive the three-hundred-dollar facility fee. The board of commissioners, the minutes record dryly, "did not view such a proposal favourably."[52] The relationship with Dalhousie was anything but painless.

The administration of the laboratory was transformed following another expansion in 1925. W.W. Kenney wrote to William Chisholm, Nova Scotia's minister of Public Works, requesting that the laboratory be placed under the direct authority of the Department of Public Health.[53] A day earlier, the hospital board concluded that it would simplify administration if responsibility for the public health work was transferred to the province.[54] The hospital and provincial authorities exchanged several different transfer schemes.[55] During the late summer and autumn of 1926, the matter was concluded.[56] The public health sections of the laboratory were hived off from the clinical ones and placed under the jurisdiction of the health department. This rationalization recognized that the public health work was rapidly growing and that it was manifestly different from the routine clinical work of the hospital. From the perspective of the VG's administrators, the transfer of responsibility also removed an expensive operation from the hospital's account books. The two laboratory services were housed under one roof but were now administered separately. There were also two discrete staff complements, with the hospital and government "each providing their own."[57] There were other minor changes to the work of the laboratory but it largely assumed a structure that would endure for decades.

The new arrangement with the provincial government, however, left Dalhousie in an uncertain position with respect to its teaching facilities. While the hospital and government arrived at an agreement, the government and university failed to negotiate any privileges. Dalhousie had enjoyed access to the facility and the laboratory director ever since the appointment in 1911 of Dr M.A. Lindsay, who was cross-appointed to the medical school.[58] Nevertheless, the relationship was never clearly defined. Dalhousie president A.S. MacKenzie described it as "loose and informal" as late as 1919.[59] It was so informal that the university and laboratory had no written agreement through the latter part of the 1920s setting out the terms of the relationship.[60] Following the reorganization of the laboratory administration, President MacKenzie lamented to the Rockefeller Foundation that, "apart from the Pathologist, who is a conjoint appointee of the Hospital Commissioners and the University, our men in Physiology, Biochemistry, Pharmacology, etc., have no access to the Hospital, there is no proper dove-tailing of the laboratory and clinical sides."[61] Clearly, this was unsatisfactory for the medical science departments, which were newly established in the 1910s and 1920s as medical education in Halifax was being reorganized.

While Dalhousie coveted enhanced access to the hospital for teaching purposes, the internal working of the laboratory was simplified by separating the public health work from the clinical work of the hospital. As 1927 began, Dr D.J. MacKenzie began working full time with the Department of Public Health.[62] Donald J. MacKenzie was appointed to Dalhousie in 1921 as a lecturer in pathology, having returned to Halifax after stints in Montreal and Baltimore sponsored by the Rockefeller Foundation. From the outset of his appointment

MacKenzie's was associated with the laboratory facilities, whether by assisting Nicholls or through an agreement with the hospital commission, "for general laboratory work."[63] In short order, his duties were expanded to include public health work and covering for Nicholls when the pathologist was on vacation.[64] Staff members remember MacKenzie fondly. "Everybody liked him, respected him and everything," recalled one worker, while another suggested he was an "absolutely marvellous character." These same workers admitted, however, that "you never got to know him that well" and that "he did have a temper ... and would let fly."[65] A colleague described MacKenzie as an "efficient teacher" and many in the faculty of medicine wanted him "back on the staff of the Medical School."[66]

The work of the public health section expanded remarkably after it was established as a separate unit. Throat swabs, VD smears, and cerebro-spinal fluid examinations increased by 33, 35, and 60 percent respectively. During 1926–27, the work of the laboratory increased by 4,676 specimens (67 percent).[67] From 1926 to 1931, the first five years the public health section was separate from other laboratory services, public health work expanded by almost three hundred percent.[68] Increases in public health work were recorded every year as the 1920s drew to a close, although about half of this work continued to centre on venereal disease testing. Other major initiatives included water testing, which increased by more than 750 percent in one year, the result of the laboratory's examining all municipal water supplies within the province at monthly intervals.[69]

THE BUREAU OF LABORATORIES, SAINT JOHN

The other major facility in the Maritimes was the Bureau of Laboratories in Saint John. This laboratory was shaped largely through the New Brunswick government's attempt to create a comprehensive public health department. As in Halifax, the New Brunswick laboratory was intended to cover the full spectrum of analyses, including clinical investigations and public health work, and even medico-legal work (not explicitly mentioned in Nova Scotia). The New Brunswick health promoters believed that the organization of the laboratory should begin with the appointment of "one of the best scientific men" available, and that this appointment "should be one of the first matters to be taken up by this Department."[70] William F. Roberts, the first health minister, was typical of reformers in the 1910s.[71] He supported suffrage, called for the licensing and regulation of restaurants and theatres, endorsed compulsory vaccination, and supported the pure-milk campaign. Roberts toured Canadian and American cities, visiting laboratories and searching for a "past master" of laboratory work.[72] Six months before the Public Health Act came into effect, the laboratory facilities were entirely remodelled and, on 18 May 1918, Dr Harry L. Abramson arrived in Saint John.[73]

Abramson was born in Russia and raised in St Joseph, Missouri. He attended Yale University Medical School and graduated in 1911. His internships took him to Connecticut and Rhode Island. Abramson earned distinction while at New York City's Bureau of Laboratories, where he worked from 1913 to 1918. At the same time, he served as the instructor in bacteriology at the Bellevue Hospital Medical College. He rose to prominence during New York's 1916 polio epidemic and his pioneering research was published in prominent journals such as *Archives of Internal Medicine, Journal of the American Medical Association, American Journal of Diseases of Children*, and the *Journal of Immunology*. There is no clear indication of why Abramson chose to leave New York.[74]

That he went to Saint John offers some indirect evidence of the international nature of laboratory work, even in a relatively small Canadian city. As with Halifax, the search for a capable director entailed contacting known authorities in the leading cities. Certainly, the opportunity to guide a provincial laboratory service held some appeal, for in New York he had worked in the shadow of the great Dr William H. Park. In Saint John, Abramson was charged with the planning and creation of a laboratory service for the entire province of New Brunswick. The laboratory was housed in the Saint John General Hospital, though the facilities were sparse. A room on the first floor was used for bacteriological work, a room in the basement for chemical work with "some additional space in the basement to be used as a preparation room." Lab animals were housed in nearby stables.[75] Several local contractors outfitted the laboratory, while equipment was procured from suppliers in Toronto and New York City. There were inevitable delays because of the scarcity brought on by war. Despite the lack of amenities and equipment, Abramson began to examine specimens on 20 May, only two days after his arrival, "through the use of the extremely meagre laboratory apparatus possessed by the General Public Hospital." Better equipment was in place by July and the laboratory began to perform several functions, including bacteriological examinations of milk and water, autogeneous vaccine preparation, pathological examination of tissues, and autopsies performed at the request of coroners. Equipment shortages delayed the planned implementation of other testing until the autumn.[76]

As in Halifax, there were no sharp divisions between laboratory and clinical medicine. Nor should the divisions between the principal general hospital and the many smaller community-based hospitals throughout New Brunswick be overemphasized. Rather, the constituent parts of the emerging health care complex need to be viewed as interconnected and interrelated. The New Brunswick health department expressed its willingness to provide assistance to other hospitals that wanted to establish small laboratories of their own to perform a "higher class of work."[77] Many community hospitals were indeed establishing such facilities. Abramson himself played a leading role in establishing a laboratory service at Moncton City Hospital in 1921–22 to perform the clinical work for the hospital and public health work for that city, including diphtheria cultures, sputum examinations for tuberculosis, Widal tests, venereal disease tests, and milk and water examinations.[78] By the 1920s, laboratories were operating under the direction of local pathologists in smaller hospitals throughout New Brunswick, including Chipman, Campbellton, Chatham, and Woodstock.[79]

These were limited facilities. Hospital administrators in Chatham and Campbellton identified their facilities as "chemical" laboratories, with Campbellton noting that tissue work was sent to the provincial laboratory in Saint John. This was by design. An important rationale for establishing a comprehensive facility in Saint John, capable of a broad spectrum of analyses, was that it would provide laboratory services to hospitals and municipalities throughout the province. The Reverend Sister Superior Walsh, the superintendent of Campbellton's Hotel Dieu, wrote to health minister Roberts that "without accessibility to proper laboratory facilities no hospital can do proper work." Nevertheless, Hotel Dieu did not have sufficient funds to equip such a service owing to the effort to rebuild the hospital following a fire in 1918. A small "chemical and pathological laboratory" serviced some needs, but specimens were regularly sent to Saint John. Sister Walsh appealed for an exemption from charges while the laboratory was put in order.[80] Roberts responded:

I quite appreciate what you say regarding the importance of the laboratory and the care of the sick at hospitals. I can also readily understand the almost equal importance of having such a laboratory ... [but to] inaugurate a laboratory that will serve all of the interests required by the general hospital at this particular time in the history of medicine would cost a great deal more money than any corporation or municipality would care about investing. The cost of the equipment itself makes such almost prohibitive, to say nothing of the money required to employ one or more that are proficient and absolutely needed to give results that can be depended upon.[81]

In New Brunswick, the laboratory service was predicated upon the idea of a central laboratory providing core services, with small hospital laboratories conducting only a limited range of routine tests. For facilities that had no laboratory of their own, the provincial laboratory would conduct any required clinical work for a nominal fee. Results from Saint John would then be telephoned or sent by wire to avoid undue delays.[82]

Ensuring that physicians could use the laboratory for their clinical work and providing timely test results was a conscious strategy to encourage familiarity with the laboratory. Many New Brunswick physicians grew accustomed to sending specimens to Saint John and receiving laboratory analyses to assist in the diagnosis and management of their clinical cases. This also ensured that the Bureau of Laboratories would obtain samples of reportable diseases. Secondly, as in Nova Scotia, through assisting physicians with their private clinical work the laboratory could surmount any lingering suspicions.

Despite the desire to provide physicians with efficient and prompt laboratory results for their clinical work, Abramson noted in his first annual report that doctors were responding slowly. He suggested that New Brunswick doctors "have not been educated" to use "a modern medical laboratory." To overcome this, Abramson authored pamphlets on diphtheria along with one on "Public Health as a Paying Investment."[83] There was also an effort to advertise the work of the laboratory to the general public through the press. Abramson frankly commented that "if patients demand the use of modern laboratory methods of precision in the diagnosis of disease, the physician will be compelled to become familiar with the uses of the laboratory."[84] Abramson also sent periodic form letters to physicians throughout the province. In 1923 he wrote that "diagnosis of diseases of metabolism is not complete without recourse to modern laboratory methods." He described the various tests performed and what they indicated. Abramson added that the laboratory "has been performing the tests aforementioned for some time and it is desirous of extending the use of this service through the Province."[85] By the early 1920s health minister Roberts suggested that the laboratory was an unqualified success.

EQUIPPED FOR SUCCESS?

The provincial governments in New Brunswick and Nova Scotia established laboratories and actively worked to ensure their usefulness for the medical profession. Yet they were small operations initially, poorly staffed and with few resources. The lack of equipment is the most obvious example of the gap between government claims and the material conditions within the laboratory. The equipment in Halifax and Saint John was very modest, but the laboratories were able to carry out the essential tests. In 1909 Halifax physicians D.A. Campbell and A.C. Hawkins reported that the equipment for urine and stomach-content examinations in Halifax was "fairly complete" and "on the whole fairly satisfactory," a lukewarm assessment at best. The microscope was in adequate repair, although the fine adjustment worked poorly, reducing its utility. The doctors also recommended the acquisition of a bell glass to protect the instrument from dust. Equipment for conducting bloodwork was incomplete, since some of it had disappeared. The apparatus for pathological histology had also vanished, with the exception of a Bausch and Lomb microtome, which was resting unused and badly rusted in a corner. Pathological specimens were found throughout the room and many were inadequately labelled. This state of affairs, while imperfect, was actually an improvement on previous inspections. A new plan was created whereby everyone except house staff was excluded from the room, which resulted in a cleaner and more orderly facility.[86]

By the time A.G. Nicholls arrived to take charge of the Pathological Institute in December 1914, the equipment was more satisfactory. The laboratory was able to complete tests for tuberculosis, diphtheria, typhoid fever, meningitis, and others, in addition to examining milk and water.[87] Specimens were usually forwarded to the laboratory through the mail. In a circular about typhoid testing, physicians were told that a small blood sample drawn from a finger-prick could be sent to the lab between clean glass-slips or even on clean white paper as long as the blood was "allowed to dry thoroughly before folding."[88] The state of specimens received at the lab varied considerably. W.H. Hattie reported in 1899 that many specimens came in "chip boxes, or on paper" and were dried out upon arrival, which made them difficult to work with and a hazard, because of "germ-laden dust [which] is set free into the air of the laboratory, thus exposing all who are compelled to breathe that air to the danger of infection."[89] Samples of pus, urine, morbid tissues, and milk often reached Hattie "in a condition which did not permit of examination."[90] In 1913 a New Ross physician sent a specimen directly to W.W. Kenney, the hospital superintendent (and not the laboratory). Kenney passed the sample along, but it was unusable. When opened, it stuck to the blotting paper and what could be used was "dry and hard." Kenney politely asked for another specimen in reporting to the physician.[91]

Ideally, samples were to be sent in clean glass bottles, which could be packed and shipped safely. Speed, the physicians were told, was of the essence, although "owing to some irregularity in the postal service" delays may occur in results coming back from the laboratory.[92] The public health officials had legitimate concerns regarding the transport of specimens. So too did postal inspectors, who had to field complaints about "bottles containing sputum and other matter offensive and dangerous to health [which are] forwarded through the mails very insecurely put up." Postal regulations banned certain materials, including explosives, "dangerous or destructive" materials or things that could damage other letters. Diseased tissues (and presumably other material), postmasters were advised, were only considered acceptable for the mail "when enclosed in specially constructed double tin cases, closely packed with absorbent matter, and with closely fitting screw caps."[93]

Few physicians followed the rules. The laboratory director took the opportunity in his annual report to remind doctors that he had a limited number of mailing cases and instructions for the collection and shipping of water samples. The problem of shipping specimens continued, however. After three decades of operation, unsatisfactory specimens were still being received. In 1931–32 the public health laboratory mailed one thousand specimen containers to physicians throughout Nova Scotia, at considerable expense. Despite this, specimens were still received in inadequate or "dangerous" containers.[94] Though the laboratory provided results only to physicians, patients often sent material to the laboratory directly.[95] The Institute of Public Health at the University of Western Ontario once received a set of teeth from a woman in St Thomas, Ontario, requesting that they be "analyzed thoroughly," for the woman suspected that they were the cause of her ill health. The lab did not perform any tests, and the woman forwarded stamps so the laboratory could "return the teeth."[96]

Despite such difficulties, physicians were unquestionably active partners in the expanding workload of laboratory facilities. Early laboratories blurred the boundaries between public health work and clinical laboratory analyses for individual physicians. This, together with the general confluence of medical interests in the Maritimes, ensured the place of the laboratory in the expanding health care complex. The careful presentation of the laboratory as an adjunct to the clinical work of the hospitals or individual practitioners assuaged any opposition to the "scientific" side of medicine. Still, the laboratory was novel and it remained unfamiliar to many physicians. As the careful instructions regarding the shipment of samples and the increasing standardization of collection methods indicate, physicians were actively participating in the expansion of the workload, but they were not fully attuned to the demands of laboratory science.

Were the early laboratories a success? Certainly, the laboratory promoters and those active in public health believed that the labs were a critical part of a modernizing health care system. As early as 1901, Dr A.P. Reid suggested that a laboratory and a qualified director were necessary infrastructure for all modern states.[97] By 1907–08, about thirteen hundred specimens were examined by the laboratory. With such a demand, the annual report wondered "how many doctors in active practice have the time to spare which is needed for these examinations, or can afford to inaugurate the expensive laboratory apparatus that is demanded?"[98] Thus, while public health was the outward face of the laboratory's work, service to the clinical work of physicians was deemed to be equally important. "The laboratory," said Nova Scotia's pioneer W.H. Hattie, "is simply intended to furnish assistance to physicians in the diagnosis of doubtful cases."[99]

It is clear that both the public health and clinical aspects of laboratory work were increasing. Laboratory analyses and the reports that were produced can be viewed as part of a system of health information that included the production of other information such as vital statistics.[100] The laboratory performed tests for diphtheria and typhoid for free, while sputum examinations for TB cost one dollar. In return, the province hoped to garner more accurate data on the prevalence of these diseases.[101] For its part, the laboratory worked to ensure quick results. Widal reactions, sputum samples, and urine samples were all examined the day they were received in the laboratory. Throat swabs were done after twenty-four hours of cultivation. With the addition of some new equipment in 1903–04, the laboratory could also examine a wide array of tissues and tumours the day they arrived.[102] The laboratory also had a vested interest in the accuracy of the results. A.P. Reid, the secretary of the board of health, produced a circular in 1896 that requested that physicians participate in determining the "accuracy" of the test for typhoid developed by Widal and Pfeiffer. The circular, which was reprinted in the *Maritime Medical News*, asked that physicians supply the laboratory with a blood sample from confirmed or suspected cases of typhoid fever, together with information on the time of onset, the severity of the attack (with the temperature), and the nature of complications. At the conclusion of the case, physicians were asked to report on the accuracy of the laboratory results and whether they assisted in a beneficial clinical outcome.[103]

Laboratory analyses, even an established method such as tuberculosis testing, were never intended to displace the clinical judgment of attending physicians. One manual of laboratory science suggested that "waiting until tubercle bacilli are found in the sputum before making the diagnosis of pulmonary tuberculosis is to jeopardize the patient's chance of recovery."[104] The laboratory worked hard to be of service to physicians in the course of their clinical work. Communications between the laboratory and community-based physicians were routine, usually consisting of sending the test results.[105] When a physician requested a test for the tubercle bacilli, diphtheria bacilli, or gonococci, they were also asked to provide the name of the patient, age, and a brief history of the case so that the laboratory could compile statistics on the incidence of these diseases.[106] Following the decision to make tissue examinations free under an initiative to address the cancer problem in 1931, the *Nova Scotia Medical Bulletin* found it necessary to remind physicians that "the giving of an accurate Diagnosis is hindered by many of the specimens arriving at the Laboratory unaccompanied by any history whatever. Often the source of the growth is omitted. A short note on the sex and age of patient, duration of tumour and other relevant points in the history of the case would be much appreciated and would be of considerable help in the giving of a fuller report on Diagnosis and Prognosis."[107] Such requests affirmed the connection between the laboratory test and clinical observation and reinforced the view that the lab analyses were meant to aid physicians.

Despite Harry Abramson's lamentations about the slow uptake of laboratory tests among physicians, it appears that many supported the laboratories. Nova Scotia's annual report for 1915–16 remarked upon the "growing appreciation" among medical practitioners "of the value of the laboratory in doubtful cases of illness."[108] The statistical evidence from both New Brunswick and Nova Scotia reveal the growth in laboratory work over time. But physicians did not abandon history taking and clinical examination. Quite the contrary, and articles often reiterated the point that laboratory tests and x-rays were adjuncts to clinical judgment, not vice versa.[109] John Harley Warner has argued that "thinking physicians who opposed the rise of laboratory science ... did not necessarily oppose science in medicine but instead objected to the new definition of what constituted science rooted more in experimental laboratory physiology than in empirical clinical observation."[110] The debate should concern not whether laboratory science was accepted by some physicians but rather whether there were competing definitions of what constituted "medical science." The science of bacteriology, because of its close affiliation with clinical diagnosis and the promise it held for therapeutics, gained the greatest acceptance among physicians. Many still questioned its relevance to the clinical setting and others expressed concern that the laboratory would displace the patient's bed as the centre of the medical universe.[111] The debate was not between "an educated elite and the average practitioner; it reflected a disagreement in values; a debate over what it was that the physician was supposed to do."[112] The problem was linking the work of the laboratory to the bedside and there were different responses.

Some endorsed the use of microscopes and laboratory fin[dings;] others viewed clinical care and the laboratory as two solitu[des. The] trajectory was clear. By 1937–38, D.J. MacKenzie reported [that, with] but one or two exceptions, every practicing physician in N[ew Brunswick] took advantage" of the laboratory service. After two short decades, laboratory work was a firmly entrenched part of health care in the Maritimes.

CONCLUSION

Laboratories were rapidly established in the Maritimes in the opening decades of the twentieth century. Large institutions, such as Halifax's Pathological Institute or Saint John's Bureau of Laboratories, were founded to serve their provinces. Local hospitals, such as the Moncton City Hospital, assumed responsibility for public health in the surrounding municipality, while other hospital laboratories simply carried out clinical work for their wards. Laboratories in the Maritimes were a response to local and provincial interests, and they were created to conduct both public health and clinical analyses. In New Brunswick, the development of a comprehensive laboratory service ensured the preeminence of Saint John, while allowing local facilities to develop and expand. In Halifax, the complex interplay among the municipal and provincial governments, Dalhousie University, and philanthropic foundations exemplifies the new place of the laboratory as a centre of health and of medical education. Indeed, the development of the laboratory was a complex blend of national and international developments and local considerations. The large laboratories in Halifax and Saint John were not bastions of scientific medicine divorced from clinical practice. Rather, the laboratory promoters fostered linkages to hospitals and physicians in smaller Maritime communities. Local suppliers played a role in constructing and outfitting the facilities, but when searching for directors the provincial departments of health made use of national and international contacts.

There was, then, a complex interplay between national and international developments and local concerns, between those in the urban centres of Halifax and Saint John and the many smaller hospitals throughout the Maritimes. A shared feature of laboratory development throughout the Maritimes was severe fiscal restraint. As a result, in large laboratories and small, the array of tests and the equipment was limited. While medical authority and provincial governments cooperated to articulate a new place for public health in the Maritimes, the result was not a final or absolute triumph for the laboratory. The tests performed were dominated by examinations of milk and water, and investigations related to tuberculosis, typhoid fever, diphtheria, and venereal diseases. While the range of tests was limited, the volume increased rapidly, requiring additions to staff and, ultimately, training programs. During this period of expansion, many young women of various backgrounds began working in laboratories, providing answers to the region's most pressing health concerns.

3

The Content of Laboratory Work

In 1920 Margaret Low left the Dalhousie medical school to begin work at the Pathological Institute.[1] She initially earned seventy dollars a month. Low was remembered as a stern person, although a "nice old soul." Relatives remember that she would on occasion bring nieces and nephews to the laboratory to play with the animals. She was also independently minded and not above raising Cain with the doctors who would come into the lab smoking. Low was allergic to tobacco smoke and "wouldn't allow it if she had anything to do with it."[2] Often, as in other facets of her working life, it was beyond her control. A woman who had been a colleague of Low's in the 1930s suggested that "she must have been there from the beginning of time."[3] Low stayed with the laboratory for twenty-seven years and there was little doubt that she was the head technician.[4]

Low was the first technical worker to be added to the laboratory staff. At the close of the nineteenth century, the laboratory was primarily engaged in public health work, analyzing sputum samples for the tubercle bacillus, throat swabs for diphtheria, and blood specimens for typhoid.[5] Public health work was the raison d'être for the laboratories in Saint John and Halifax, though both performed significant amounts of clinical work, as argued in the previous chapter. While the range of tests that were performed remained small, there was a significant trend to ward more and more tests. In 1896–97, the first year for which figures are available, the laboratory performed tests on 266 specimens.[6] Through the first decades of the twentieth century the laboratory built a steady clientele and the health department suggested that the lab's equipment and personnel were being "taxed to [the] utmost."[7] By 1920 the specimen total had exceeded 3,300, and it topped 6,200 in 1921.[8] Moreover, a new public health age was dawning in Halifax, expressed most dramatically by the expansion of Dalhousie's medical school. By 1919–20 the work had become too much for one individual, and for the first time there were additions to the staff. More than anything, however, the work was defined by the burden of disease, which, together with increased attention to public health and a period of hospital construction and expansion, framed the conditions under which laboratory work developed and expanded in the Maritimes, and elsewhere.

LABORATORY TESTS AND THE BURDEN OF DISEASE

By the 1920s, the laboratory had largely fulfilled the promise of medical bacteriology that began with Robert Koch four decades before. Bacteriology revealed the agents of disease, hidden in milk and water or carried by individuals in their bloodstream. Medical bacteriology had developed precise tests to identify and differentiate the micro-organisms of various diseases, including the Widal test for typhoid, the Schick test for diphtheria and the Wassermann test for syphilis. Laboratories were helping physicians determine diagnoses. Laboratory products, including serums and vaccines, provided physicians and public health officials with a means to combat certain diseases. Health departments now had a fighting chance to control the outbreak or spread of some diseases. The laboratory had a critical role to play, offering both the state and clinicians several definitive tests to aid diagnosis and protect the public health.

An illustrative example of the nature of laboratory work is the Widal test, a common analysis for typhoid in the formative years of Maritime laboratories.[9] To complete the test, a small drop of the patient's blood would be placed on a glass slide. Four drops of water were added and the slide was tipped from side to side to ensure that the cells were hemolyzed. Four drops of formalinized[10] culture of typhoid bacilli were then added and the slide tilted again to mix the solutions. If the reaction was positive, agglutination would occur in about one minute. If the results were doubtful, the excess fluid was drawn from the slide, which was then dried and fixed over a flame. The dried slide was then stained with methylene blue and examined under a low-power microscope lens. When the test was positive, the bacilli were seen in large clumps. In this early and relatively straightforward example of a laboratory test that was common throughout the 1910s, 1920s, and 1930s, the laboratory worker had to be adept at working with the slides, preparing the formalin and carefully staining the slide if the test was to be useful to the clinician.

Though Widals were common they paled in comparison with the enormous volume of other analyses, such as sputum samples for tuberculosis. Testing sputums grew dramatically from the First World War to the mid-1930s at the Pathological Institute. Tests increased from 359 in 1914-15 to 984 in 1921-22 and 1,463 by 1928-29. These numbers continued to grow, reaching more than 2,100 by the end of the 1920s, almost 4,800 in the depths of the depression, and topping 8,000 samples by 1935-36.[11] Clearly, examining sputum was an important task in the early laboratories. Patients were given clean, wide-mouthed bottles and a new cork. The sample was collected from material coughed from the lungs (not, for example, secretions from the nose). Once obtained, the sputum was placed in a Petri dish or on blotter paper. The worker then removed any larger particles in the sample and set them aside. These particles were then squeezed out with another glass slide until the layer was thin and even, covering about half of a slide. The resulting smear was fixed over a flame and the specimen destroyed. Heating was necessary because it overcame the tubercle bacilli's resistance to staining. Once stained, the bacillus retained colour when washed with alcohol and mineral acids. Next, carbon fuschsin was poured on the sample. The worker held the dry end of the slide and heated the end covered by the stain until steam appeared. After three minutes, the stain was rinsed with water and the smear decolourized with acid alcohol until it turned pinkish grey when washed again with water. The smear was then counterstained with Loeffler's methylene blue for thirty seconds. Again the smear was washed, then dried with blotting paper and heat. These procedures demanded the careful attention of the worker. One they were complete, the sample was ready for examination.

With a stain such as Ziehl-Neelson, common in 1930, rodlike bacilli cells appeared red, while the cells were blue when examined under a microscope. The smear was viewed by carefully moving the slide until the entire area was examined. Ideally, a worker spent as many as five minutes looking at a smear. When several of the telltale rods were identified, a positive diagnosis could be made. But identification required a degree of proficiency. As with other tests, accuracy was dependent upon the acumen of the workers. Smears that were too thick might retain stain even when the tubercle bacilli were not present. Small imperfections in the glass slide could also trap the red stain. It was only with practice and skill that the observer could identify the bacilli. Picking out the larger particles required patience. Other errors could occur. Burning the sputum when fixing the sample, boiling the stain instead of steaming it, allowing the stain to dry off, or decolourizing for too long all reduced the sample's utility. Time was another issue. The laboratory manuals may have suggested five minutes per slide, but in a busy laboratory, as Nicholson and other authors acknowledged, thirty seconds was probably more typical. If the technique was poor, early cases could easily slip detection. The laboratory worker had to be diligent and conscientious if sample results were to be satisfactory. Even when everything was done correctly, the work did not end. Workers had to confirm positive findings by examining other specimens. In suspected cases of tuberculosis, even a negative sample did not lighten the workload. Further tests were required because the bacilli might not be present at all times.

Diphtheria was among the early diseases identified in the laboratory and one of the first to be targeted with a biological agent developed in the laboratory. Behring and Kitasato laid the foundations for serotherapy by proving that immunity to diphtheria, induced in a guinea pig, could be transferred passively to another animal. They identified the substance in the serum of immune animals and dubbed it "antitoxin."[12] With antitoxin, physicians and public health officials could defend the populace against diphtheria. By the beginning of the First World War, the Nova Scotia Department of Public Health was procuring diphtheria antitoxin and distributing it both to local health boards and (at a slightly greater cost) to individual physicians.[13]

Diphtheria also provides another useful illustration of the work of the laboratory. The Schick test, introduced in 1913, was very reliable. Only individuals showing a positive reaction were susceptible to diphtheria. A minute amount of diphtheria toxin was injected intradermally and, in people who were susceptible, a red area developed. Those who enjoyed immunity did not develop this redness. The Schick test began in the laboratory. The diphtheria toxin, which had to be fresh, was stored in a refrigerator and diluted immediately before injection. The laboratory worker broke off both ends of a capillary tube, being careful not to lose any of the contents. One end of the tube was placed into a rubber bulb, which was squeezed to mix the contents with five cc's of saline. The saline was drawn up from its vial and expelled several times to rinse out the toxin, and the resulting dilution was subsequently corked. The worker then turned the vial several times, shaking the contents. The diluted toxin was only good for twenty-four hours. A very small amount, exactly one-tenth of a cc of the dilution, was then injected between the layers of skin on the surface of the forearm. When done correctly, a small white weal appeared on the skin. When done incorrectly (if the fluid went too deep or was lost on the surface of the skin), another injection was necessary.[14]

Diphtheria also provided clear evidence of the utility of the laboratory for establishing diagnoses. During the spring of 1916, diphtheria was reported in several areas of Nova Scotia, and it reappeared in the fall. Not surprisingly in the midst of war, there was also some concern for the troops who were afflicted with the disease. The outbreak was mild and the mortality was reported to be quite low. However, the Nova Scotia Department of Public Health noted "amongst soldiers, a large proportion of the cases lacked the membrane and other signs usually regarded as distinctive of diphtheria, the diagnosis being made by laboratory methods."[15] Clinical observation could be misleading and the department reminded readers that specimens should be forwarded to the laboratory in all doubtful cases. Only in this way could the effort against diphtheria be assured of progress against often hidden threats to health.

The laboratory proved its utility in a different way when diphtheria broke out at the Victoria General Hospital in 1916.[16] Two suspected cases were quarantined until two separate laboratory tests each showed them to be negative. It is unclear what motivated the multiple trials, but it is nevertheless significant. For the two women who were confined, the negative laboratory tests were sufficient to spring them from quarantine. The hospital took extraordinary measures to combat the outbreak. The isolation building quickly filled, and the hospital segregated identified "carriers" in the general wards from other patients. The hospital also imposed strict limits on who would be admitted. Nurses, maids, and even patients were subjected to throat swabs, as the laboratory endeavoured to identify potential cases before the disease spread. As Naomi Rogers has suggested, germs were often identified with filth but could appear "insidiously in those who seemed clean and healthy."[17] Indeed, by the end of January, two hundred and twenty cultures for diphtheria had been completed, and thirty-three carriers identified among the patients and workers. These individuals showed no clinical symptoms and could have prolonged the outbreak, possibly with devastating consequences.[18]

The health of the province's largest hospital was one matter, but whole towns were at risk because of impure water supplies. In 1902 W.H. Hattie reported that inadequate sewage disposal and impure water were contributing to a generalized problem of typhoid fever.[19] Again the next year, the Provincial Board of Health's annual report noted solemnly that while typhoid fever continued to be prevalent, it was so "insidious that it does not rouse popular excitement." It was well known that the route of transmission was infection of the water supply by improper sewage disposal, and the report concluded that every water supply should be tested to ensure purity.[20] The laboratory, as with the 1916 diphtheria outbreak in the hospital, would render the invisible menace visible, and water testing became a major focus of the laboratory's work.

As the examples of diphtheria and water testing illustrate, the laboratory responded to changing demands placed on it. A decade after its founding, for example, tissue samples from animals were a notable feature of the laboratory's work, reportedly the result of increasing public pressure for pure milk and an unadulterated food supply.[21] The effort to ensure a safe milk supply, steeped in political and social considerations, was also a fight in which the laboratory could assume a central place. Earlier efforts to ensure safe milk centred on inspecting the cleanliness of cow herds and dairy farms and preventing the adulteration of milk. In 1908, the Canadian Medical Association (CMA) milk commission recommended higher standards for milk production, handling, and storage. The CMA embarked on a public campaign to advertise that milk carried many diseases, such as typhoid, diphtheria, and tuberculosis.[22] Provincial health authorities in Nova Scotia agreed, suggesting that the milk supply was a significant contributor to Halifax's "excessive" infant mortality rate.[23]

The crusade for clean milk was a major health issue in all urban areas throughout North America. The *Maritime Medical News* commented in 1906 that "what were formerly spoken of as water-borne diseases may now be more fairly, perhaps, termed milk-borne." The editorial also suggested a connection between the safety of the milk supply and rural water. "The water supply of cities is now fairly looked after ... But milk is obtained from the farms, and the water supply of farms is, as we know, not always above suspicion – and indeed in many cases is very bad indeed."[24] This is an interesting example of rural conditions being portrayed as unhealthful, in contrast to urban conditions, based on the evidence of the laboratory. Rural settings may look better, but laboratory investigation could reveal the "truth." Paul Bator has boldly argued that Toronto's successful campaign for clean milk resulted in the imposition of urban standards on farms, standards that ultimately undermined rural claims of a healthful environment. The resulting regulation of milk signified the emerging dominance of cities, both legally and symbolically.[25]

The efforts of reformers did not go unchallenged. For example, William F. Roberts, New Brunswick's first minister of health, introduced pasteurization to Saint John in 1923. This prompted a hue and cry from local dairies, which accused him of having a financial stake in forcing milk to be pasteurized. Some of his opponents even left sour milk on his doorstep. In part because he supported unpopular legislation such as pasteurization, Roberts was defeated in the 1925 provincial election.[26] Laboratory tests yielded empirical evidence of hitherto hidden threats, while providing "scientific" evidence of unhealthy rural environs. Such evidence served an important ideological function in an urbanizing Canada.

There is no question that the trend was toward greater regulation of the milk supply. But a nuanced analysis is required. As in the case of diphtheria when suspected carriers escaped quarantine, milk dealers could occasionally reap the benefits of an accurate test that proved their milk was clean. But a laboratory test could not remove the damage of suspected sources of infection. Reporting to the provincial secretary in 1904, Dr A.P. Reid recounted a visit he made to a dairy farm near Milford that was prohibited from selling milk in Halifax. Reid found the claim to be unwarranted. The operator of the farm had her business ruined, and "prejudice prevented her, and still prevents her from being able to resume her wonted business."[27] For a dairy, negative publicity was detrimental and could happen at any time, with disastrous results for the proprietor. Reid, incidentally, suggested a scheme of licensing milk vendors as a protection not only for the public health but also for vendors wrongly suspected of unclean operations. A clean bill from the laboratory could prevent false accusations but could not undo them once made.

The hospital diphtheria outbreak and the question of milk testing are important reminders that the laboratory's influence extended beyond the confines of public health or clinical medicine. While the work of the laboratory consisted of analyzing bits of people or of goods, the test results had profound social impacts. A positive venereal disease test could reveal infidelity. A negative diphtheria test could see the removal of a quarantine sign from a household and the return of the family members to work, school, or other neighbourhood activities. A milk analysis that proved unfavourable could ruin a small family dairy. In 1950 Albert Deutsch highlighted the importance of accurate laboratory reports in an article for *Woman's Home Companion*, where he wrote that "many a reputation has been ruined, many a marriage destroyed, many a life shortened, many a savings account wiped out by erroneous laboratory reports on syphilis and gonorrhea."[28] The laboratory may have been hidden from view, but it was hardly isolated from the public. Laboratory reports were harnessed to identify and define the limits of a broad range of social "problems," including those of immigrants, inadequate services in rural settings, or the dangers of city living.

VENEREAL DISEASE

There is no better illustration of the confluence of social concerns and laboratory work than efforts to control venereal disease (VD). The Canadian campaign against VD began in earnest in 1917. Prominent reformers, among them Dr C.K. Clarke, reported that evidence from Toronto General Hospital revealed a syphilis rate of between twelve and thirteen percent. Gonorrhea, while not assigned a percentage, was argued to be the cause of half the cases of sterility in women and a quarter of cases of blindness in children. Syphilis caused miscarriage or stillbirth, insanity, and congenital deformities and generally exacted a heavy toll. Canada, reformers argued, needed legislation.[29] Dr P.H. Bryce, the federal health gadfly, was convinced and he added the VD problem to his list of reasons for establishing a federal department of health.[30] The Academy of Medicine, a Toronto medical society, framed the necessary measures to control VD. The academy reported that both a medical program and an education program for physicians were required. With respect to the former, any general hospital in receipt of federal money should participate in diagnosing and treating venereal diseases. They also recommended that the federal government grant participating provinces ten thousand dollars for laboratory costs.[31] Most provinces passed legislation that addressed VD in the years before 1920.[32]

Venereal disease control was one of the ten divisions of the federal health department when it was established in 1919. VD control was a budget of $200,000, the second largest in the department, was devoted to controlling the "secret plague."[33] This money was designed to assist the provinces, and the federal government itself was to be minimally involved in delivering the program.[34] The federal government supplied the funds in proportion to provincial population, while the province delivered the services. The effort to diagnose and control VD thus became an early example of federal conditional grants to the provinces.[35] Federal-provincial conferences were held in February and May 1919, and resolutions guiding the disbursement of federal funds were approved. In addition to establishing free clinics and beds in hospitals, treating inmates of provincial institutions, and establishing a provincial VD division with a specialist in charge, provinces had to maintain diagnostic laboratories to be eligible for grants.[36] In Halifax, the laboratory conducted syphilis tests free of charge. The Department of Public Health's annual report for 1919–20 noted simply that "the laboratory is being used for [venereal disease] tests to a much greater extent than formerly."[37] Following World War I, Nova Scotia also established treatment facilities in Halifax, Sydney, and New Glasgow and procured equipment for centres in Yarmouth, Lunenburg, and Amherst.[38] This expansion was funded on a fifty-fifty basis with the federal government, tangible evidence of Ottawa's interest in the matter of venereal disease.

The success of the campaign for a robust response to venereal disease stemmed in part from its definition as a social rather than a health problem.[39] The story of venereal disease in Canada is well documented and it need not be recounted here. It is interesting to note, however, that the historiography largely replicates this focus on the social relations of VD; historians have thoroughly analyzed the information campaign but have paid scant attention to diagnosis or treatment.[40] Even studies that address the campaign to conduct Wassermann testing fail to consider *who* was conducting the tests.[41] Laboratory workers remain utterly invisible. Margaret Low was one such worker. The first addition to the staff of Nova Scotia's Pathological Institute, she was appointed chiefly to handle the large volume of venereal disease tests and paid through a special VD account. Low was the frontline worker assisting in the effort to provide free diagnosis and treatment to those suffering from the disease.

Testing for VD testing became a steady part of the laboratory's work through the 1920s. In Halifax, statistics were reported for the first time in 1919–20, showing that 610 tests had been conducted that year. The numbers grew through the 1920s, peaking at 2,564 in 1921–22 and staying in that range until the Wassermann was displaced by the Kahn test. The change in the testing protocol did not slow the growth in work. The number of syphilis tests conducted in the laboratory increased sixteenfold between 1919 and the mid-1930s.[42] While the work of the laboratory expanded over the first two decades of the twentieth century, testing for venereal diseases enhanced the pace of activity. The volume of testing vastly increased and ultimately the public health work, of which VD testing was a substantial part, was established as a separate service. Staff were added to handle the increased number of samples, which required care and diligence.

The Wassermann test, introduced in the Halifax laboratory in 1915, offers an illustration.[43] Acute syphilis promotes antibodies in the blood as a reaction against the infection. In the test tube, these antibodies will bind together in the presence of an antigen. After the reaction, the bound complement can no longer join with hemolytic serum to hemolyze the sensitized red cells that are later added. In other words, when syphilitic antibodies are present in the serum, the red cells will sink to the bottom of the test tube, leaving only a clear fluid above. If the blood sample is free of syphilis, the complement will join with the serum and cause hemolysis, resulting in a bright red fluid in the test tube. This reaction, known as complement fixation, could also be used for other bacterial diseases using different antigens. Venipuncture, a common technique to draw blood for many tests, was used to obtain the sample. A tourniquet was applied to the patient above the elbow. If the veins were small, they were dilated by opening and closing the hand, slapping the vein vigorously, having the patient swing the arm, or even immersing it in a hot-water bath. Once the vein became accessible, the patient was ready to be punctured. The thumb of the left hand was placed adjacent to the vein to steady it and the needle slowly inserted, ideally at a twenty-degree angle. Five to ten cc's of blood were drawn, the tourniquet removed, and the needle withdrawn.

Once obtained, the sample was sent to the provincial laboratory within twenty-four hours and without too much agitation. Temperature extremes were also to be avoided. If it took longer than two days to transport the sample to its destination, it was better to send serum only. In such cases, the blood sample stood for three to six hours in a warm room until the clot began to contract. The test tube was spun in a centrifuge to drive the clot to the bottom and one or two cc's of clear serum were drawn off and placed in a new vial with a cork stopper. Daniel Nicholson, who wrote a popular laboratory manual in 1930, warned that "every extra step increases the hazard of contamination."[44] With samples arriving in Halifax or Saint John from all over the Maritimes, the extra work was necessary. Without it, the samples became hemolyzed and therefore useless.

Workers had to ensure that the pipette and vial were clean, dry, and sterile. The equipment of the laboratory had to be thoroughly cleaned with soap and water and then rinsed three times in distilled water. The equipment was then dried and sterilized in an "ordinary cooking oven."[45] Indeed, cleaning slides, coverslips, or test tubes was a major feature of laboratory work. Ordinary soap and water, followed by rinsing and wiping, sufficed for simple cleaning. For the more stubborn stains, Daniel Nicholson suggested "grit soap" like Bon Ami. More extensive cleaning was, of course, frequently required in the laboratory setting. New glassware had to be cleaned with a seventy-six to ninety percent solution of alcohol, with one percent acid. The glass was soaked for five minutes then rinsed, preferably with distilled water. If the glassware contained infectious material, the worker boiled it in a two percent solution of sodium carbonate for fifteen to thirty minutes, then rinsed it. If there were stubborn stains on the glass, the worker applied heat, then an acid solution and then immersed in another cleaning solution overnight. Some of the solutions, including a sulfuric acid bichromate mixture, were very caustic to the skin. Not surprisingly, given the harshness of the material and the work required, Nicholson noted that it was "questionable whether the time and breakage involved to clean coverslips that have been covered with balsam and oil, makes it worth while or whether it is better to discard them."[46] The busy laboratory worker was probably thankful for that concession. In the later 1930s, Nova Scotia's public health laboratory director estimated that on a typical day, between one thousand and twelve hundred items were "boiled, cleaned sorted and sterilized" in a small room inadequate for such a large task.[47]

Wassermann tests, while routine, were considered to be highly technical. Nicholson suggested that the test should only be conducted "at a large laboratory where a competent serologist is giving special attention to the reaction."[48] Sufficient controls needed to be in place to ensure the accuracy of the result, but accuracy depended upon specimen collection and patient determinants as well. Hemolyzed blood or the presence of chemicals or bacteria (due to inadequate cleaning) limited the accuracy of results. Moreover, even in a syphilitic patient, the serum examined could yield different results. Remissions and exacerbations in syphilis patients are common and the Wassermann test was positive only when the disease was active. High fevers or anesthesia could also produce false positive results, as could the presence of other diseases, including nonpulmonary tuberculosis, diabetes, or cancer. The work may have been precise and the test complex, but the interpretation of the results was uncertain at best. Despite these very real limitations, a positive Wassermann was of more diagnostic value than a negative one. If a positive reaction resulted, the test was typically repeated. If it was again positive, the presence of syphilis was almost certain.[49] The Wassermann test was positive in fully ninety percent of syphilis cases and gave the laboratory enormous diagnostic authority.

POLIO

The laboratory exposed health threats, whether in the city or the country, in people or in foodstuffs. As the diagnosing of diphtheria, milk and water testing, and venereal disease analyses demonstrate, the public health laboratory, its tests and workers, revealed hidden perils that threatened the well-being of the populace. The volume of laboratory work expanded accordingly and entirely new kinds of work were added to the routine. Vaccines were first manufactured in Nova Scotia in 1908, and the annual report for that year anticipated "a great increase in this work."[50] While this aspect of laboratory work did not become a major component of local activities, the production of convalescent serum in response to the polio scare of the 1930s provides a good illustration of this feature of the work.[51]

Polio had long attracted the attention of scientists. No less a figure than Simon Flexner, the director of the Rockefeller Institute for Medical Research, put the newly opened Rockefeller Hospital in New York on the trail of polio in the spring of 1911. Three of the hospital residents, Francis Peabody, George Draper, and Alphonse Dochez, began their investigations.[52] Research was also conducted in various other laboratories, but the answers remained elusive. There were competing theories of transmission. Flies were thought to be culpable, while others thought that dust or other fomites could spread the disease.[53] Uncertainty over transmission did not diminish confidence in the laboratory. By the second decade of the twentieth century, after all, bacteriology and immunology had combined to provide a host of products to combat disease. Antitoxins were developed for diphtheria, rabies, and tetanus, while vaccines existed for plague, cholera, yellow fever, and typhoid. The laboratory also offered a variety of diagnostic techniques for diseases such as tuberculosis, diphtheria, typhoid, and several venereal diseases. In 1909 the polio virus was identified and, following the course of other discoveries, clinicians and the public alike believed that the laboratory would soon provide a way to identify those infected and, more importantly, a treatment. One of the earliest attempts was an antipolio serum made of the blood of patients who had recovered from the disease.[54] Two decades later "convalescent serum" stood as the only treatment available for polio victims.

Preparing convalescent serum required the blood of patients, ideally those who had been free of the disease for between a week and a month. The blood was collected in large test tubes under aseptic conditions. An eighteen- or twenty-gauge needle was used and the drawn blood was placed in cold storage overnight. The next day, the clot was loosened from the side of the tube and the sample spun in the centrifuge for thirty minutes. The serum was drawn off using a pipette and the worker had to ensure that no hemoglobin was present in the sample. If present, hemoglobin would generate a severe reaction when injected intraspinally. With this precaution, the sample was tested for sterility. If sterile, it was diluted and could be stored for public health use for up to a year. Many public health officials in the late 1920s thought that convalescent serum was valuable for polio cases in the preparalytic stage. The dose would be given intravenously when the patient first presented and the diagnosis was established. This was followed by two intraspinal injections of between fifteen and twenty cc's with a one-day interval. Smaller doses were given to young children. The laboratory seemingly offered public health officials a response to the dread disease.

In Halifax interest in polio was intense.[55] In 1929 the public health laboratory prepared convalescent serum for two cases of infantile paralysis.[56] It was the first use of the serum but it foreshadowed a growing battle with the polio menace. Convalescent serum was explained to be a "remedy" for the disease.[57] The laboratory's annual report for 1929–30 confidently asserted that "when given early it is apparently uniform in its results." It reprinted data from Ontario, which claimed that when given on the first day of illness, the serum prevented paralysis in one hundred percent of cases, while those receiving its benefits on the second day had an eighty-seven percent chance of recovering without paralysis. The Department of Health compiled a list of donors within Halifax and prepared and stored serum in the laboratory. Outside of the city, lists of donors were also prepared, and the services of D.J. MacKenzie were offered "to any hospital centre, for technical advice and assistance, in the preparation of the serum locally."[58] The laboratory advertised in the *Nova Scotia Medical Bulletin* that physicians should "telephone or wire the Laboratory as soon as cases, or even suspected cases of Infantile Paralysis are discovered" to obtain serum.[59] Perhaps in an effort to expedite this process, the next year the laboratory prepared a large amount of serum and distributed it to medical health officers throughout the province.[60]

Doubts were first raised about the treatment in the 1931–32 annual report.[61] The preparation and distribution of the serum had come to be a major component of the work of the laboratory. But was it a successful enterprise? In the early 1930s the threat of polio ebbed once again, with only sporadic occurrences and no epidemics. Reports of infantile paralysis were "disturbing" to parents and health officials alike, and the sequelae, notably paralysis of the limbs, was "dreaded." Public health officials in Nova Scotia believed that some of the dread was removed because of the use of convalescent serum. They did, however, acknowledge the controversy. "While doubt has been cast, by some, on the efficacy of the Serum," the annual report for 1933–34 stated, "consistently good results have been reported from quarters where it has been extensively used."[62] A slight shift in the discourse of the treatment protocol accompanied this acknowledgment, for the serum stockpile was now viewed as "an emergency precaution," in marked contrast to the policy of just a few years before when it had been widely distributed. Subsequent reports continued the trend: in 1935–36 it was noted that the serum was "still being used" and distributed "following the custom of late years," and the report for 1936–37 saw fit to mention "differences of opinion regarding the effectiveness of convalescent serum," while stating blandly that "nevertheless it was used." Another treatment regimen, a nasal spray, was reported for the first time, even though its trial in Toronto met with disappointing results.[63]

OTHER LABORATORY WORK

In addition to the large issues, there were other changes to the work of the laboratory. The blood transfusion service was further developed in

1921 and, following the discovery of insulin, a blood chemistry service was established in 1922, consisting of a modest range of tests.[64] Tissue examination had always been an important part of the work of the laboratory, but it received a significant boost in 1931 when the exams were made free in Nova Scotia's effort to address the "cancer problem" and through the establishment of the Victoria General Hospital's Cancer Clinic the following year.[65] The free examination of tissues was touted as "the beginning of an effort to attack the Cancer problem," which the health department believed would result in earlier diagnosis and treatment.[66]

To meet the expanding workload, laboratory directors frequently requested more staff. In Halifax workers of varying backgrounds, most of whom were women, joined Low and Nicholls. Dr Harry L. Abramson foreshadowed the need for staff at the Bureau of Laboratories in the 1918 annual report, the first one. Abramson indicated the need for chemical apparatus and a chemist to assist in pathological work and analyses of milk, water, and food. He also suggested the need for a "well trained man" to assist in the work and suggested that such a person could be hired for fifteen hundred dollars per year.[67] The need for additional staff in Saint John became a familiar lament, together with yearly demands for increased space and apparatus. A chemist was added in 1921 but lacked adequate equipment. Abramson noted wryly that "we have the chemist, but nothing for her to work with."[68] Cecelia LeBrun worked for only about six weeks, leaving in 1922 to return to her native Sydney, Nova Scotia.[69] Other workers assumed her duties. Though specialized, the work was routine and it was not unusual for a co-worker to fill in during vacations, illness, or in the wake of resignations. Laboratory staff may have preferred a particular aspect of benchwork, even undertaking special courses, but they were expected to be familiar with the full range of analyses that were performed.

Despite the ability of workers to adapt to new responsibilities or learn new procedures, Abramson felt that the laboratory could not keep expanding its work with such a limited staff. With no additions forthcoming, the Saint John facility tried to lighten the increasing workload in other ways. In the mid-1920s, for example, it delegated milk testing to laboratories in Moncton and Campbellton, in an effort both to improve the accuracy of the analyses and limit the volume of routine work in the Bureau of Laboratories.[70] As in the Halifax facility, there was an increase in the number of tests performed over time, determined both by innovation and the burden of disease. Venereal disease testing, so important in the early work of the laboratory, was delayed thanks to the lack of equipment.[71] Kahn tests joined the Wassermanns in 1925 and for a period both tests were performed on any blood sample received by the laboratory.[72] Increased testing, however, was not equated with an actual increase in the incidence of venereal disease. The laboratory characteristically reported that the increase was the result of heavier use of the laboratory by physicians. William Warwick, New Brunswick's chief medical officer, proffered another opinion. In the depths of the depression, Warwick noted, people were seeking treatment in the government clinics in greater numbers, thereby inflating the figures.[73] While serology tests were free for New Brunswickers, they also generated income for the laboratory. In 1928–29, for example, Prince Edward Island, without a public health laboratory of its own, paid the Bureau of Laboratories $285 to perform Kahn and Wassermann tests on its behalf.[74]

While public health tests such as those for venereal diseases garnered the laboratory the attention of the federal and provincial governments and additional funding, routine clinical tests such as blood counts were equally important for shaping the workload of the laboratory. To complete a blood count, a large drop of blood would be obtained from a patient. Workers would visually count the number of cells in selected squares (four if they were counting leukocytes and five for red cells). In the case of leukocytes, the count was multiplied by fifty and with red cells, four zeros were added to yield the number of cells per cubic millimetre. The resulting counts were then compared against normal values. Differential white cell counts were even more demanding. A blood sample was then divided into three segments and counts made in each one. One hundred white cells were counted in each segment and the kind of cells tallied to calculate the percentage of each variety. Lymphocytes, monocytes, eosinophils, and neutrophils were all enumerated and compared with normal ranges. An increased count of these white cells indicated different things. A greater percentage of eosinophils, for example, could indicate a recent allergic reaction, whereas heightened levels of neutrophils or lymphocytes could indicate a bacterial or viral infection, respectively. Suffice it to say that the counts were critical and these manual differential counts, known as "manual diffs," were both time-consuming and demanding.

Urinalyses provide another example of routine laboratory work and were conducted with basic equipment in nearly every hospital. Sugar tests and albumin tests were frequently requested and a number of steps were required. The laboratory worker heated water and while it was warming, the specimens were numbered with a marking pencil. Every test tube or slide was assigned a corresponding number. The worker shook the specimen vigorously to mix any sediment, and filled a fifteen cc centrifuge tube. The sample was then spun for between four and five minutes. Next, 0.5 cc's of urine and five cc's of another solution were mixed and when the water reached a boil, the tube was immersed for two to three minutes, after which it was allowed to cool. The laboratory worker recorded the odour, colour, clarity or turbidity, and specific gravity of the sample. The albumin test was then performed and the sample left to stand while the sediment was examined microscopically. Finally, all the findings of this protocol were dutifully recorded. The work was routine, insofar as it would be repeated countless times in a typical day and hundreds of times a year. The routine took the worker through seven precise steps that allowed for the efficient examination of urine, with little loss of time.[75]

Even routine tests were shaped by shifts in clinical knowledge and the burden of disease. Urinalyses, which had been performed regularly at the Bureau of Laboratories, increased by eleven hundred analyses during 1922-23. As insulin treatment took hold in the Saint John General Hospital, the diabetic patients strained the laboratory's capacity, for they required daily examination of their urine for sugar levels. The treatment of this patient population also led to the establishment of a blood chemistry service the same year. The necessary equipment cost three thousand dollars.[76] Nevertheless it was thought to be a judicious expenditure, allowing patients to avoid travelling to other metropolitan centres and physicians to treat and manage their patients with greater precision. By 1935, the Bureau of Laboratories conducted "all the routine laboratory work of a modern general hospital, a provincial tissue diagnostic service, the provincial public health and medico-legal work and the administration of a serum depot handling large quantities of expense and perishable prophylactic, therapeutic and diagnostic biological reagents."[77] As with the Pathological Institute in Halifax, the Bureau of Laboratories in Saint John fulfilled many functions for the health care complex. The scope of work performed in emerging Canadian laboratories challenges the assumption that these were places with highly discrete roles, separate from other dimensions of clinical care, medical education, or public health work. Several aspects of the emerging health care system intersected in laboratories such as those in Halifax, Saint John, and elsewhere.

With such a variety of tests and rapid growth, the familiar calls for more staff, equipment, and space echoed through the Bureau of Laboratories' annual reports from the beginning. In an interesting argument, Abramson suggested that the limited space in the Saint John General Hospital made accurate work difficult, noting that an efficient laboratory depended upon accurate work. The lack of space was also blamed for curtailing the laboratory's utility, for the staff could not perform all the tests demanded of them. Finally, Abramson raised problems associated with venereal disease testing. Patients refused to be seen in such a public setting. Often physicians would not send their patients for venereal disease testing, to save them possible embarrassment. A private consulting room was necessary, so that people could report for venereal disease testing "without the fear of [other] persons knowing or surmising" why they were at the laboratory.[78]

In addition to more and better-designed space, there was a need for more equipment. There was only one centrifuge, perhaps the most important piece of equipment in any early laboratory. Should the centrifuge break, serology and milk testing would stop. There was also a need for new equipment to keep up with developments in blood chemistry.[79] The second annual report, conjuring the historic rivalry between two port cities, suggested that if "Nova Scotia can afford to have a laboratory building, New Brunswick is not so poor but that she can make the necessary expenditure."[80] Thereafter, the chief medical officer and the laboratory director made the lack of space and equipment a regular feature of their respective annual reports. Little would change through the 1920s as economic collapse consumed the reform impulse. Finally, when a new Saint John General Hospital was announced in 1928, the plans included increased space for the laboratory.[81] On 1 October 1931, the new laboratory was opened, although quickly Abramson noted that the new space was just adequate.[82]

THE EXPANDING WORKLOAD

Despite the persistent complaints about equipment, space, and staff, laboratory tests grew to become an established part of public health and clinical care during the 1920s in many Maritime centres. In Halifax, for example, the number of tests increased from 759 in 1914–15 to more than to 8,753 a decade later and exceeded twelve thousand tests by the end of the 1920s.[83] It became tenable to suggest that medical examinations were "incomplete" without laboratory results.[84] Tests could not, however, provide answers to every question. When reporting interesting or unexpected results to physicians, particularly those in conflict with the clinical opinion, the laboratory director asked to be updated on developments in the case.[85] In early 1920, a woman was admitted to the Victoria General Hospital for an ulcerated leg, but while in hospital she "developed Typhoid symptoms." A letter to the Massachusetts-Halifax Health Commission (MHHC), which was paying for the patient's care, stated that the diagnosis was typhoid, despite the negative laboratory findings.[86] Another case sponsored by the MHHC saw the patient investigated by the full array of diagnostic tests, including x-rays and a variety of laboratory tests, among them Wassermann, blood examination, and testing of cerebro-spinal fluids and globulin. None of these shed "any light on the condition" and the patient died.[87] Testing also yielded more favourable results. An Africville resident was admitted to the Victoria General on 3 January 1924 and subjected to a "thorough examination, including gastric analysis, blood examinations, blood Wasserman [sic], Barium series and examination of stools." He was diagnosed as having a gastric ulcer and improved enough to be discharged (at his own request) on 24 January.[88]

Related to the issue of uncertainty in diagnoses was the question of error and standards. Recalling the implementation of the Wassermann tests in Halifax, D.J. MacKenzie suggested that "almost every laboratory used its own modification, usually in the direction of over simplification."[89] Error in observation was, of course, nothing new in medicine.[90] Nineteenth-century instruments such as microscopes, thermometers, and sphygmographs all provided suspect information and even fine instruments could yield different interpretations when used by different individuals. As the laboratory gained acceptance, it too came under increasing scrutiny. Nineteenth-century chemists and physiologists recognized that differences in patterns of work, emotion, digestion, or even the weather could influence test outcomes. Specimens, as outlined before, arrived in various forms affecting the reliability of tests. Laboratory media were not exempt either. Guinea pig serum used in Wassermann testing or media for bacteria cultures could vary considerably from laboratory to laboratory, or within one lab. The considerable diversity in experience and training, described in subsequent chapters, also heightened the potential for variation and laboratory error. Stanley Reiser argues that the seriousness of laboratory error was compounded by the confidence physicians placed in the results, a confidence that "often blinded [physicians] to the errors caused by carelessness in collecting, preserving and transporting the material to be examined."[91]

False diagnoses affected not only the public health of a municipality but also the private lives of its citizens. If a laboratory cleared an individual who was, in fact, infectious, an outbreak could ensue. Incorrect positive diagnoses could prove inconvenient or disastrous for the patient, particularly (but not restricted to) tests for venereal diseases, as suggested earlier. These considerations led the Boston health department to initiate a groundbreaking study to determine the accuracy of laboratory work in 1919. Specimens were sent to fourteen laboratories, which were asked to examine for gonorrhea, syphilis, diphtheria, tuberculosis, and typhoid fever. The resulting data showed wide discrepancies in the testing of identical specimens. This investigation, together with the generally poor reputation of many laboratories, led American organizations of pathologists, bacteriologists, and chemists to request

that the American Medical Association (AMA) begin to supe
facilities. In 1924 the AMA's Council on Medical Education a.
tals, with the cooperation of expert opinion, established stand
issued an approved list of facilities. The results of this, which w
emulated in Canada, were that "physicians began to urge that ...oora-
tory technicians pass licensing examinations" and that "laboratories
formerly run by technicians came under the control of physicians,
while some of the worst commercial laboratories went out of business."[92] The ultimate impact was the consolidation of laboratory work
in hospitals or other government institutions and rising workloads.

By the late 1920s, laboratory staff in Saint John and Halifax were working at a furious pace to keep up with the demand for tests. The public health section in Halifax was working on a regular basis with some twenty hospitals throughout Nova Scotia.[93] In New Brunswick, personnel from Saint John were loaned to the Moncton Hospital and Victoria Public Hospital in Fredericton, and special courses were given to workers from Fredericton, Woodstock, and St Stephen.[94] In Nova Scotia, D.J. MacKenzie requested the services of a part-time technician shortly after his appointment as director of the public health section.[95] MacKenzie even noted in one annual report that "while his staff is underpaid and often overworked ... they are always ready to continue working even on public holidays in order that reports may be sent out promptly." Efficiency in the laboratory depended, to a large extent, on the training and experience of staff members, but the material benefits of their work did not accrue to the laboratory workers. MacKenzie expressed the opinion that it was difficult to maintain an adequate staff complement in the Halifax laboratory when other facilities offered workers twice the salary.[96] To remedy this, MacKenzie suggested regularly scheduled increments up to a fixed maximum. He argued further that the work was "exacting and dangerous and that the laboratory's efficiency depended upon the training and experience of every member of the staff."[97] The following year, the situation had not improved. Work continued to increase and the staff was "taxed to the utmost." In response, the provincial health officer recommended the immediate addition of at least one new worker.[98] Five years later, the laboratory was still understaffed, resulting in workers putting in "considerable overtime." Despite facing this "extra burden," the staff managed to examine specimens and report on them promptly.[99]

As the 1930s dawned, the laboratory was well-established and the demands placed upon it continued to grow each year.[100] By this time the laboratory service was an integral component of the health department.[101] It played a significant role not only in disease control but also in prevention. A couple of years later, the Department of Health would again assert that "[a] properly equipped, efficiently staffed, and well organized public health laboratory is, without a doubt, one of the foundations of modern health work. It is an important factor in coordinating all health activities. In providing an accurate and prompt diagnostic service, the laboratory helps to maintain a satisfactory relationship between the practicing physicians and the health department."[102] The workload was increasing both on the bacteriological and pathological sides, but especially the former. The services were "hives of activity" that presumably translated into "better methods of disease control."[103] Nova Scotia's health department reported near the end of the 1930s that the large number and variety of specimens the laboratory received suggested the frequency with which physicians in practice and public health officials were using the facilities. The expanding workload was cited as evidence of a growing interest in prevention and an increasing commitment to providing "service to all the people."[104]

CONCLUSION

Multiple levels of government, the medical school, the burden of disease, and a changing orientation among practicing physicians and public health officials alike combined to shape the context of laboratory work through the 1920s. The workload expanded to meet these interests. The laboratory was gaining credibility as a site of education, for improving the public health, and assisting attending physicians in their clinical endeavours. There were successes that bolstered this credibility. The efforts against diphtheria and in the diagnosis of tuberculosis and venereal diseases were shining examples of the utility of laboratory work. Milk and water testing became established parts of municipal and provincial public health campaigns. Even when answers remained elusive, as in the effort against polio, the laboratory occupied a central place, indicative of the new commitment to the investigative enterprise.

The work of the laboratory was complex, as illustrated in the descriptions of such common tests as the Widal, Schick, and Wassermann. For each test, preparing the samples required care and precision. The mounting volumes of samples and tests through the 1920s also meant that laboratory workers had to work quickly and with precision. While laboratory manuals such as Daniel Nicholson's 1930 *Laboratory Medicine* suggested five minutes per slide for sputums, the pace at the bench moved much more swiftly. While the preparation and testing of samples clearly involved skilled tasks, many of the tests were routine. The laboratory worker was also responsible for mundane but important chores such as cleaning glassware.

From the beginning, the laboratory work was filled with ambiguity. It was a combination of skilled jobs and mundane work. Some tests were complicated, others routine. The 1920s and 1930s were characterized by a dramatic growth in the scope and content of laboratory work, as new areas such as blood chemistry developed. Once the laboratory was firmly established for both public health and clinical work, staff numbers were increased. Laboratory workers grew from a few individuals in the early 1920s to an important service within the hospital and the emerging health care complex. The growth of laboratories and the expanding workforce, however, must not be assumed to be evidence of the increasing and linear trend toward specialization. Rather, as subsequent chapters reveal, the workers were incorporated in a way that served a variety of interests, much like the facilities in which they worked.

4

Not Just Bench Warmers: Labour in the Laboratory

In the United States in 1875, there were 661 hospitals, a number that had increased to slightly over two thousand by 1900. From 1900 to 1929, hospitals were established at the rate of two hundred per year.[1] Although no annual data on hospitals were collected in Canada before 1932, Ontario experienced a similar expansion. The number of public general hospitals in that province increased from fifty-one in 1900 to 139 in 1930.[2] Hospitals were founded throughout the Maritimes in these same years, to augment the few that were built in the late nineteenth century. In the dangerous mining district ruled over by the Dominion Coal Company, for example, the need for health services was profound in the early twentieth century and two company houses near the No. 2 colliery were converted to hospitals. In eastern Nova Scotia, the Sisters of Saint Martha, a congregation founded in 1900, staffed many of the new facilities.[3] The Sisters entered hospital work in 1902, first at St Joseph's Hospital in Glace Bay. Their work quickly expanded. They ministered to the sick and performed a wide array of tasks in the newly established St Martha's Hospital, founded in 1906 in a small home on West Street in Antigonish that could accommodate only six patients.[4] The Sisters also staffed St Rita Hospital (the Ross Hospital) in Sydney beginning in 1920 and St Mary's Hospital in Inverness in 1925.[5] Hospitals were also established in other Maritime communities. Yarmouth's hospital could trace its origins to 1912[6] and Western Kings in the Annapolis Valley to 1922.[7] In New Brunswick, the Moncton Hospital opened on King Street in 1902, St Joseph's was established in 1914 to service Saint John's Catholic population, and Miramichi Hospital was founded in Newcastle a year later.[8] There were also a number of specialty institutions in the region.[9] Mere blocks from the Victoria General Hospital in Halifax stood the Children's Hospital (founded in 1909) and the Grace Maternity Hospital (1922).

The Sisters of Charity began operating the Halifax Infirmary in 1886, while the Halifax Visiting Dispensary serviced the "deserving poor" of Halifax as of 1855 and opened a Dartmouth branch in 1877.[10] There were other hospitals, including an infectious disease hospital, several military hospitals, and private facilities.[11]

This period of development and expansion created employment opportunities, as hospitals became increasingly complex. The place of the hospital as a site of medical and nursing education has long been acknowledged and hospitals predictably began to train other workers to staff the emerging diagnostic departments around the Maritimes. By the 1920s, hospitals everywhere were becoming specialized, at least superficially. Mary Kinnear, using an appropriate metaphor, described the Canadian hospital of this period as "a modern, scientific, progressive laboratory where doctors could perform their new skilled techniques."[12] A commemorative booklet produced for St Martha's Hospital in Antigonish during a 1925 fundraising campaign stated as much. "In the hospital alone," the booklet declared with optimism, "can a complete and accurate diagnosis be made. Here alone can be found the x-ray machinery, the laboratories, the individual records so necessary to a useful diagnosis."[13] Stanley Reiser has described this complexity as a "partnership" by the 1930s, with doctors organized to practise medicine, while pharmacists, nurses, and laboratory technicians "were integrated toward patient care."[14] There was, however, a very clear hierarchy in the hospital and the interests of some workers were subordinated to those of physicians and administrators.

To meet the demand for enhanced diagnostic services, hospitals needed to train staff to fill positions at the laboratory bench, in the x-ray room, or in other settings. Laboratory workers were recruited from a wide range of academic backgrounds. Some had university science degrees, while many moved directly from secondary school to their jobs in the laboratory. Evidence from the Maritimes demonstrates that there was no one route to the laboratory bench, although many of the women obtained university degrees or at least some university experience. Regardless of their background, workers received basic training from laboratory directors to meet the needs of a particular setting. Workers refined their skills through on-the-job training, either through tutorials or supervised practice. These workers, in turn, trained others who followed.[15]

The demand for laboratory personnel was strong in the 1920s because of the creation of laboratories to meet hospital accreditation standards and the growing range of public health tests. But there was not a ready pool of workers to fill the demand. When the Massachusetts-Halifax Health Commission searched for a laboratory assistant and a technician in 1920, for example, a suitable candidate could not be found.[16] To meet their labour requirements, hospitals across Canada initiated training programs. Beginning in the 1920s, the Pathological Institute trained workers for positions throughout Nova Scotia and, occasionally, elsewhere. D.J. MacKenzie recalled that the first workers (other than those who would assume duties in the Pathological Institute) were trained in 1923. In that year there were two students, one from Glace Bay and one from St Martha's Hospital in Antigonish.[17] In his 1923 report, A.G. Nicholls wrote that "there have been several applications from persons desirous of qualifying in laboratory work to be taken into the department as students." While the laboratory could not accommodate all the requests, two students did receive a course of instruction. This was repeated the following year, with two more students taking an eight-month course in "laboratory technique," both of whom found work in other institutions. Nicholls recognized that the demand for laboratory technicians was increasing and that his own facility would require more staff when the expansion was complete.[18] The annual reports make no further reference to training students in laboratory technique, but the practice likely continued for Nova Scotian hospitals. In New Brunswick, the Bureau of Laboratories similarly trained workers to assume positions in hospitals throughout the province.[19] Many of the earliest hospital workers to take up duties in laboratories, particularly in smaller institutions, were nurses.

NURSES

Nurses, as Margarete Sandelowski has illustrated, were absolutely key to the "scientific and technological transformation of health care and medicine" that took place in the early twentieth century.[20] The addition of new diagnostic services or other departments created a demand for competent and capable staff but the limited financial resources meant that hospitals needed inexpensive options.[21] One strategy was to employ new workers who would assume duties in multiple departments. In many settings, however, the more common strategy was to add the responsibility for these new departments to the workday of nurses. Susan Reverby has argued that beginning in the 1910s, as hospitals added new services and departments, nurses could be found working as laboratory or x-ray technicians, social workers or physiotherapists.[22] Sandelowski describes the nurses in these settings as "assistants"[23] but they were often in charge of these areas. When Prince Edward Island appointed its first laboratory technician in the early 1930s, there was widespread agreement that she would work alone.

Nurses were a flexible labour pool for hospitals and could often be found at work in many different departments, as the example of the x-ray department at Halifax's Victoria General Hospital illustrates. Dr William H. Eagar was appointed roentgenologist at the Victoria General in November 1919.[24] At the time, he was promised that the hospital would supply "an assistant, not necessarily a so called technician, who shall be an employee of the nursing department." The nurse, when working in the x-ray plant, was under Eagar's direction, who assumed responsibility "to train and develop" the prospect.[25] An early candidate was Miss Kathleen Sullivan, who lived across the harbour in Dartmouth. Superintendent W.W. Kenney informed her that training would last for six months and that during this time there would be no salary or allowance, although he did suggest that meals might be had for free at the hospital. Moreover, training would not guarantee Sullivan the position.[26] She declined this rather unexciting offer and in late January 1920, Eagar plucked a senior male nurse from the staff to train as a "technician."[27]

Under Eagar's direction x-ray work expanded. Between 1920 and 1924 the volume of work in the x-ray service, which was established as a department in May 1921, doubled. As the workload expanded, the divide between the manual work of preparing x-rays and the diagnostic work grew. Physicians retained their position as interpreters of data in the diagnosis and management of their patients. They could consult with their colleague, Dr Eagar, if a case was particularly vexing. Meanwhile, the work of the x-ray plant, including the maintenance and preparation of equipment, would be left to staff members who had nothing to do with interpreting the images. The separation of the manual work from the intellectual was assured and through this division, the pre-eminence of physicians was guaranteed. It would become a familiar pattern, not only in the x-ray services around the Maritimes but also in the laboratories.

The increasing demand for laboratory tests prompted hospital administrators at the Victoria General to establish a training program. Students often made their way into the laboratory of Dr A.G. Nicholls, who trained two at a time, although requests sometimes exceeded this number.[28] The length of training varied considerably. Nicholls reported in 1924 that two women received a course of eight months' duration.[29] Dhirendra Verma argued that the training was intended for high-school graduates with neither university experience nor nursing education.[30] The historical evidence suggests otherwise. Nurses were among the earliest students,[31] though they typically trained for only a few weeks. All received practical education at the bench and frequently received instruction in both laboratory and x-ray work.

Nurses from rural hospitals recalled their laboratory experience in oral histories collected during the 1980s.[32] Greta MacPherson was born in Glace Bay on 14 February 1902. Her father ran a grocery store in Donkin, six miles away, and only came home on weekends. Greta was twenty when she entered Glace Bay General, after teaching school for one year. The hospital superintendent, Isabel MacNeil, had been a neighbour of young Greta's and that influenced her decision to enter training at the general hospital. Greta began in the summer of 1922 and went to work performing x-ray and laboratory work upon her graduation. A staff nurse had performed this work, after taking a "short course" in Halifax. It was MacNeil who suggested that MacPherson take the same course. The new nursing graduate was not enthused and several other young women had previously refused the offer.

MacPherson demonstrated some aptitude for the work, though she worried that these responsibilities would interfere with her nursing. The superintendent reassured her that there was not much lab work, only blood counts and urines, and that the x-ray work would not be burdensome. MacPherson recalled that the superintendent "was anxious for me to do this. Nobody would touch it. So I took instruction." There were few places available in most hospitals for graduate nurses, but MacPherson was offered a position doing lab work, x-rays, and some staff nursing; she even administered anesthesia. She spent ten years doing this work and grew to enjoy it, particularly the x-ray work. MacPherson stated that "to be a nurse-technician, you're a jump ahead of when you've just taken a technician's course because you already know how to handle patients."[33] The combined skill-set likely secured MacPherson a position, but her ability to work throughout the small hospital, which did not have the patient volume to justify hiring dedicated staff members for each department, benefited the hospital as well. Other local hospitals, eager to offer patients all the amenities of modern medicine, made similar demands on their nursing staff.

Another Cape Bretoner, Flora K. McDonald, recounted a similar experience. McDonald was born on 1 August 1909. Her father worked at Dominion Coal in a variety of mining jobs and her mother worked in a store prior to marriage. McDonald entered the nursing school in Glace Bay on 15 September 1928 and graduated three years later. She remembered that there was a superintendent and an assistant but that student nurses staffed the hospital almost exclusively. McDonald recalled that there was "a lab technician and an x-ray technician – a lab and x-ray, and when we went in first they did them both."[34] Dorothy Allan, who began at the Yarmouth Hospital School of Nursing in September 1931, recalled that there were only four registered nurses there, including the night supervisor and the operating-room nurse. One of the nurses also served as the laboratory technician, according to Allan.[35]

Clara MacKinnon was born on 1 January 1910, the youngest of four children. Her father was a baker and her mother a teacher who, although she did not work following her marriage, did offer instruction in English to Glace Bay's substantial immigrant community. MacKinnon described her parents as early socialists who believed in education. She did not want to be a nurse but was nevertheless enrolled in nursing school by her mother. Her mother then wrote Clara to inform her that she had been accepted and was to enter on 12 October 1929. She had not been asked. MacKinnon, recounting this story, suggested that her mother had always wanted to be a nurse. MacKinnon had been living in Michigan with a married sister, and when they "motored home" Clara returned to Glace Bay. She entered the training school at the Glace Bay General Hospital, and her education was typical of nurses. Doctors offered instruction in surgery, medicine, obstetrics, and pediatrics. The assistant superintendent taught nursing procedures and techniques, the superintendent taught ethics, while the "lab and x-ray technician" offered instruction in bacteriology. Following graduation, MacKinnon worked in the x-ray department, where she had spent some time during training. The nurse who was in charge took three months off to study for an examination, so MacKinnon was placed in charge. She also noted that all the nurses could perform the basic blood examinations and urine analyses required in the laboratory.

The expectation that MacKinnon would perform a variety of tasks reflected the attitude of the superintendent, who believed that nurses should be able to "do everything, every department of the hospital and fill in," recalled MacKinnon, "whether it's the kitchen ... even the furnace room ... and you should know the lab, the x-ray, the pharmacy, the laundry, the whole business. Make yourself aware [so] that you know that you can fill in anywhere if you're in the hospital ... Jack-of-all-trades."[36] The experience of other nurses adds weight to this account. "You were everywhere," one recalled, "and you could handle anything."[37] Evelyn Purdy, who graduated from the Yarmouth Hospital School of Nursing in 1920, worked on the hospital books for one month, spent a month in the kitchen assisting the cook, and undertook x-ray training.[38] Clearly, the superintendents of smaller hospitals wanted nurses who could fulfill a variety of roles within the hospital. George Weir, in his landmark survey of nursing education in Canada, thought the demands were growing large enough and the hospital complex enough that nurses would need more training in the future.[39]

Nurses, then, worked in many departments in the modernizing hospital. In Halifax, the *Echo* noted in early 1925 that many smaller hospitals throughout Nova Scotia now had their own x-ray plants and that the medical profession was making increasing use of the equipment on a regular basis.[40] Nurses often staffed these early x-ray plants. The expansion of diagnostic services created demand for training that was broad, brief, and practically oriented.[41] When a request arrived from Colchester County Hospital in 1925 for a combined course of instruction, Kenney noted that "during the last few years we have in a few instances given a period of instructions [*sic*] in these subjects to nurses from other institutions ... Attempts have been made to comply with their wishes, but we have never felt that the best could be accomplished in this dull way, [and] now the process is strongly discouraged at least."[42] Hospital superintendents generally wanted combined x-ray and laboratory courses. But nurses occasionally exercised some choice in their continuing education. Nurse MacKinnon, for example, arrived somewhat unexpectedly in Halifax to receive a course in x-ray and laboratory work. After two weeks in Halifax, Kenney wrote to Florence Merlin, the superintendent of New Waterford's hospital, indicating that MacKinnon wanted to continue her education in laboratory technique and wished to pursue her x-ray training the following year.[43]

The courses for nurses were short and the x-ray and laboratory services were committed to providing instruction free of charge. Kenney informed Eagar that it was the policy of the hospital's management board that the "specialties at least, should serve and aid in the development of smaller hospitals of the Province to the fullest extent."[44] In correspondence with Nicholls, Kenney wrote that the management board believed the laboratory building should be "freely" opened to smaller hospitals in the province for instruction in laboratory technique.[45] "Freely" referred to access and not to cost. The Victoria General's board ruled that "so far as the hospital equipment for the work was concerned, it should be available for teaching purposes for other Provincial Hospitals free of charge ... [but] the Pathologist and Roentgenologist would be open to negotiate with the applying institution as to a reasonable fee for their services."[46] Eagar was the first to raise the issue of compensation for his training efforts and Kenney supported his request, provided that the arrangements were not a "hindrance" to those seeking instruction.[47]

It is clear that many nurses were at work in the expanding service departments of Maritime hospitals. There they played an integral role in the scientific and technological transformation of the hospital. By the

1920s and 1930s, nurses in laboratory and x-ray departments were applying new technologies to the diagnosis, management, and, in the case of x-rays, treatment of patients. Nurses' use of technology, and the technology itself, were placed in the service of physicians. In this way, the work of nurses in laboratories can be viewed as part of a broader history of technology in health care. As Sandelowski recently wrote, doctors viewed nurses "much like stethoscopes and surgical instruments, as physical or bodily extensions of physicians."[48] In the new scientific hospital of the twentieth century, patients were routinely given blood tests or asked for urine samples and physicians now depended upon others to collect and analyze the samples. When a North Carolina hospital faced a shortage of interns before World War 1, a local physician suggested that tasks such as urinalyses, blood counts, and medical histories be delegated instead to nurses.[49] "In order to harness the benefits of this new technology," Sandelowski argued, "physicians had to share its use with nurse (and eventually also with a host of new technicians whose jobs were created in response to it)."[50]

Thus, there was a clear expectation that nurses would perform a variety of tasks throughout the hospital as a normal part of their duties. However, the nurses did not derive the same cultural or professional authority from technology that physicians did. Indeed, as Rosemary Stevens has argued, such broad responsibilities helped to define nurses as "all-purpose female service workers without a defined monopoly of scientific skills."[51] Physicians used stethoscopes to diagnose disease; nurses used them to collect information that was then passed to the physician for interpretation. Nurses' engagement with technology in the diagnostic services, rather than providing a fresh impetus to professional claims, served instead to blur their role, while concurrently confirming their subordinate position to physicians.

It is not entirely clear how nurses themselves felt about their use of technology and their new duties within the hospital. Lavinia Dock, a leading American nurse, recognized that opportunities in the service departments could alleviate some of the overcrowding that was characteristic in American nursing as early as the 1890s. Dock suggested that departments such as dietetics or pharmacy were promising employment alternatives for nurses. Moreover, Dock believed such services would be better served by being staffed with nurses.[52] Rank-and-file nurses expressed ambivalence toward these new roles. Some nurses, such as Greta MacPherson, feared that new duties would interfere with their nursing work. Greta, after all, suggested that nobody else wanted to do the work, evidence that such positions were not desired.

"LOCAL GIRLS"

The expansion of diagnostic service departments placed new demands on nurses who worked in smaller hospitals. The growth of these departments also introduced entirely new workers. Through the 1920s, a broader range of tests became standard and the volume of tests increased significantly. A greater public health effort, the demand for milk and water tests, and the enhanced clinical use of laboratory tests solidified the position of laboratory science within medicine and created more jobs at the laboratory bench. To staff their own facilities and provincial hospitals, the larger laboratories in Saint John and Halifax initiated training programs in the 1920s.

One of the earliest additions to the staff of the Halifax laboratory came through the sponsorship of the Massachusetts-Halifax Health Commission (MHHC). The MHHC was incorporated in May 1919 and financed with a quarter of a million dollars that were left over from donations made to Halifax in the wake of the Halifax explosion.[53] The MHHC had a broad mandate to pursue public health work in the city. As part of that effort, it financed the addition of another pathologist and a technician to the laboratory staff. However, the MHHC failed to consult with the hospital authorities. The incident is indicative of the commission's aggressive pursuit of its goals. It also indirectly reveals the ambiguous place of the laboratory even in the medical and public health community. B. Franklin Royer, the head of the MHHC, was fully informed of most aspects of public health work in Halifax. But clearly he did not understand that the laboratory was under the jurisdiction of the hospital board.

The hospital did agree to the new staff, after ensuring that the full salary of the additions would be the responsibility of the MHHC.[54] In 1921 the commission appointed Dr Harry D. Morse to the public health laboratory as a "laboratory assistant" to deal with samples originating from the MHHC's health centres, and to make "an extended series of observations on the milk and water supply of the City of Halifax."[55] Morse resigned in May 1922 to assume a fellowship in urology.[56] In October 1923 the commission appointed Dr Foster Murray to the laboratory staff.[57]

The MHHC also paid for a technician, Gertrude Hines, who began work in 1920-21. She had "considerable laboratory experience,"[58] including five years' experience in England. Hines initially worked for seventy dollars per month, which was increased to ninety the following summer. She had a brother who taught at Bloomfield School in Halifax, which likely made the transition to her new city easier.[59] Familial complications also informed Hines's decision to leave the following spring, when she returned to England following her father's death.[60]

Another early addition was Dr Margaret Chase. Chase was appointed as a *technician* in May 1923 in Dalhousie's Department of Pathology and Bacteriology, and Nicholls also wanted her as an assistant in the Pathological Institute.[61] In correspondence, Nicholls noted that Chase, who had graduated the previous day, was "willing to come on a technician's salary, that is $75 per month." Chase did not labour in the laboratory for long. Her subsequent career took her to the United States, including Philadelphia and New York. Indeed, Dr Margaret Chase may be said to be the first female physician to be employed by the Victoria General Hospital, a distinction traditionally granted to Dr Eliza Perley Brison.[62] That a physician would toil in the laboratory as a technician suggests the many obstacles facing women doctors in establishing a practice.

Margaret Chase was the thirty-fourth woman to graduate in medicine from Dalhousie and if the careers of her predecessors are any indication, it was very difficult for a woman to establish a practice in the Halifax area.[63] Many of the women graduates left to pursue careers in the United States. Others, including Annie Hamilton, the first woman to earn her medical degree (1894), Florence O'Donnell (1901), Martha Philp (1902), and Grace Rice (1903) went to China. Some pursued medical missionary work in India, including Blanche Munro (1904), her classmate Jemima MacKenzie, Minnie Spencer (1910), and Bessie Thurrott (1922). Florence Murray, who rendered exemplary service during the Halifax explosion and the Spanish flu epidemic in the city, could not build a successful practice in the city and left for Korea in 1921.[64]

Very few female graduates took up practice in Halifax, although Annie Hamilton spent nine years in the city's north end before departing for China and Grace Rice established a successful practice on Spring Garden Road. Graduates did find opportunities elsewhere in the province, including Chester, Bridgewater, Sydney, and Dartmouth. Occasionally familial relations could facilitate entry into practice, as they did for Katherine MacKay (1895) and Clara May Olding (1896). Despite long years of preparation, marriage meant the end of a medical career for other women, including Dr Florence O'Donnell, who stopped practising upon her marriage and dropped the appellation "Doctor" as well. For others, success in medical practice was an elusive prospect. Eliza MacKenzie (1904) opened her office in Charlottetown, Prince Edward Island, but could not build a successful practice. She left for New York, pursued nurses' training at St Luke's Hospital, and continued to work as a nurse until her death in 1930.

Others did enjoy a measure of success. Eliza Perley Brison (1911) was a pioneer in Nova Scotian psychiatry, serving as the superintendent for the Home for Feeble-Minded Girls from 1918–25. Elizabeth Kirkpatrick (1915) enjoyed a lengthy career in the United States breaking new ground in psychoanalysis, before returning to Halifax to join the faculty of medicine in 1960. Ella Hopgood (1920) established the St John Ambulance in Nova Scotia and was appointed the assistant superintendent of the Nova Scotia Hospital in 1928. The pattern that saw women graduates of Dalhousie's medical faculty work in institutional settings replicated those found elsewhere and distinguished women doctors from their male colleagues who could be found most often in private practice.[65]

Opportunities within medical practice were deeply gendered and shaped different career paths for men and women, even if they worked within the same locale or profession, or graduated from the same medical school. Women physicians could be found at work in the "feminine specialties," which in the nineteenth century meant obstetrics and gynecology and in the twentieth, pediatrics, public health, teaching, and counselling.[66] Certainly, this pattern was apparent among Dalhousie graduates in medicine and serves as an important reminder that gender was embedded in the organization and provision of health care. As health care services were expanded through the early decades of the twentieth century, hospitals and agencies alike turned to women to fill the new positions. Women physicians, who were unquestionably privileged when compared with other women, did not enjoy the same opportunities as their male colleagues. For women workers in Maritime hospitals and health agencies who lacked the status (however constrained) of a medical degree, the contours of gender politics shaped the addition of new duties to old ones or the creation of entirely new job opportunities.[67]

By 1924 A.G. Nicholls was growing increasingly dissatisfied with staffing levels at the Pathological Institute. He requested "more help in order to be a better service," despite the additions of Hines, Chase, and the others. By that time, the laboratory staff consisted of two technicians and one stenographer. The laboratory was undergoing further expansion in 1924, and the hospital board would not consider Nicholls's request for more staff until the renovations were complete. Instead they suggested that Dr D.J. MacKenzie and Margaret Low, both of whom were doing public health work, could assist Nicholls in the hospital work for a modest increase in their salaries.[68] Thus even a worker with a seemingly well defined role, such as Low, did not perform a narrow range of work in the emerging and ever-changing environment of the laboratory. In late 1924 Low's expertise was pressed into service for the hospital, for which she received an additional fifty dollars per month.[69] Low's employment by the hospital, primarily to cut sections for microscopy, was to be temporary. There is every indication that she would have continued to perform this double service, except that she was off duty between 18 March and 25 August 1925. When Low returned, she worked exclusively in the public health laboratory, and Deborah Henderson was subsequently hired to do the histological work, at the same rate of fifty dollars per month.[70]

MacKenzie and Low continued to assist with the hospital work until the laboratory staff underwent a thorough reorganization.[71] When the new laboratory was complete, in the late fall of 1925, the matter of expanding the laboratory staff was again considered. Nicholls presented his needs at a meeting of the board of commissioners in November 1925. In addition to the current workers, Margaret Low and Albert Baker, two additional technicians would be added. With only a "limited" education, Baker had difficulty working with percentages. Therefore, he required supervision when preparing the solutions or media for the laboratory work. He was, however, "fairly capable" at sectioning and fixating tissues and was good at staining sections and smears. This, together with occasional work in blood chemistry, was the range of Baker's work.[72]

One of the staff to be added on a permanent basis was Deborah Henderson, who was appointed at seventy-five dollars a month.[73] Like Low before her, Henderson was known to Nicholls and enjoyed a Dalhousie connection. Beginning in September 1924, she assisted in Nicholls's bacteriology course, earning thirty dollars a month. Nicholls had designs for Henderson, hoping "to train her fully in course of time as a technician."[74] In addition to his tutelage, Nicholls encouraged her to enroll in a biology course and arranged a tuition exemption from the university. Nicholls also requested that Henderson be paid for eight months, one month longer than Dalhousie desired. He asked that she be paid for May, noting that her family was not very well off.[75] By the fall of 1926, Henderson was described as "a mainstay in ... sectioning tissues and preparing tissue slides." She was also permitted to take two afternoons a week in order to pursue a class in histology.[76] She worked at the laboratory until 31 January 1931.[77]

Henderson's experience provides a clear illustration of the important ways that the laboratory director exerted influence over employment, wages, and additional training. For selected workers, opportunities for further training were accommodated and encouraged. Nicholls recruited a promising young student to the bench through personal knowledge of her skills. He also took her family situation into consideration when determining her period of work. Such details, unimportant in isolation, demonstrate how family, education, social networks, and other factors were intertwined in the making of a labour force. Abstract ideals about skill or educational preparation, themselves socially constructed, did not operate apart from these concerns.

In addition to Henderson, Nicholls had a young man from Scotland in mind, whom he contemplated employing at the rate of eighty dollars a month. An "office girl" would also be employed and paid by the provincial public health department.[78] By 1926 the pathological service had Nicholls, Baker, and Henderson and a student assistant, "who is also the interne [sic] for ambulance calls, and who has classes which must be attended." The intern performed the "simpler clinical work for the hospital" such as examination of stomach contents, assisting in blood chemistry, urinalysis, and examination of smears. There was a half-time stenographer, although this service was irregular and sometimes unavailable. Finally, the assistant professor of bacteriology was traditionally appointed the assistant in the pathology laboratory (though without salary).[79]

Some of the workers, like Henderson or Low, had earlier connections with Dalhousie University. Many others who worked in the laboratory had university degrees. Patricia S. Tingley, who held a BSc with special training in chemistry, joined the laboratory in 1926–27.[80] Peggy Cameron was the daughter of a railway engineer. She was educated in Stellarton and New Glasgow, before enrolling in Dalhousie in 1926, at age eighteen. Cameron completed her BA in 1930, though she returned to university for additional courses.[81] She received an appointment as a part-time technician in June 1929 and was "a volunteer worker" during her last year of university. Cameron took further courses in bacteriology and biology in June 1931, after which she was made a full-time technician. Her work in these courses complemented her earlier studies of chemistry, and biochemistry, and in the opinion of the director, "she will be exceptionally well trained for Public Health Laboratory work."[82] Another "well-trained" worker was Mary MacDonald, appointed in June 1938. Like Cameron, MacDonald had pursued university classes in bacteriology, chemistry, and biochemistry, in addition to physiology. She had also worked as a "voluntary technician at the laboratory for nine months" before her appointment.[83] Other workers had graduate education, including Katherine Miller (MSc) and Pauline Webster (MA).[84]

More typical was Rose Phillips. She completed several chemistry courses and many of the premedical courses in biology and zoology. Further training on the job augmented her formal education.[85] In 1933, for example, Phillips was selected to pursue "special instruction" in the radium department, so that she could supply in the event of illness or during vacation periods.[86] Such learning on the job, always a component of the laboratory worker's life, was also recounted by other workers. Edna Williams recalled that "we all learned there on the job ... Everybody just worked together and learned ... from whoever happened to be doing the job at the time."[87] A worker from the 1930s recalled how "there were always people who would come in to take the course." This same worker did not feel it was much of a training program, however. "Usually it was somebody who wanted to get away from home everyday. It really wasn't very much of a course. And they helped out, you see, they were another body to do things. And they would pick up quite a bit." She also thought that the students were overwhelmingly "local girls." She remembered that "some were doctor's daughters, or somebody would have an interest or want during the summer something to do or it was a change perhaps from going down to work in a store, or they didn't want to become a nurse or who weren't interested in school teaching. So this was a big deal, I suppose. So we always had somebody."[88]

Many of the workers volunteered in the laboratory before being appointed to the paid staff. In Saint John, the laboratory frequently benefited from the presence of "several unpaid assistants or learners" without whom increases in the volume and nature of work would not have been possible.[89] Indeed, the unpaid labour force in the Bureau of Laboratories usually numbered about five or six.[90] One such volunteer was Moncton native Sylvie Comeau, who worked without pay in various laboratory sections during 1936 and 1937.[91] Not surprisingly, volunteers from the laboratory would often leave to pursue paid work when opportunities became available.[92] Occasionally, unpaid workers like Peggy Cameron or Mary MacDonald would volunteer for one year and then be added to the staff of the laboratory the next.[93] Cathy Arnold, became an "assistant technician" in Saint John midway through 1936–37 and was paid three hundred dollars. By 1943 Arnold was the senior technician in the facility and earned almost a thousand dollars annually. Helen Frances was, like Arnold, a volunteer in the mid-1930s, before joining the paid staff. In 1936–37, she earned only eighty-three dollars for her services, before being appointed as an assistant the next year, earning the full annual wage of six hundred dollars.[94] Rachel Hunter, although similarly described as a volunteer, did earn some money before a resignation opened up a position for her, suggesting that voluntarism in the laboratory was not a strict category. Like Arnold, Hunter would eventually rise to the position of senior technician in 1943.[95]

As the 1930s dawned, the director, seven technicians, and a caretaker staffed the Bureau of Laboratories in Saint John.[96] After more than a decade in operation, there was little indication that the pace of work would slow. Between 1930 and 1933, the total number of examinations increased by more than one hundred percent. This was, in part, precipitated by an agreement between the Saint John General Hospital and the Bureau of Laboratories, which saw the bureau perform all clinical laboratory work for the hospital. In the same three-year period, hospital work increased by an astronomical 234 percent. This suggests how important laboratory tests were becoming for the hospital patient and the clinician. Other work also increased by eighteen percent, making this a period of remarkable growth.[97] However, only two additional staff members were added to the complement and, the Medical Officer's annual report confessed, "both of these at totally inadequate remuneration for the painstaking work required of them."[98]

With the expanding workload, there was a recognized need for specially trained laboratory workers both to staff the Saint John laboratory and to service the increasing needs of hospitals throughout New Brunswick.[99] Gertrude Marks, the earliest identified "laboratory assistant" in Saint John, earned one thousand dollars a year initially. She spent over a decade at the bench, retiring on 1 December 1929. During 1924–25, she was made the "Chief Assistant," when both her experience and the addition of several other workers made such a designation meaningful. This brought with it a salary increase from $1,350 to $1,500 per annum.[100]

As in Halifax, there were others at work in the facility, besides Marks and Abramson. Wayne Mobley provided caretaker services and other duties for the laboratory, while Nettie McNamara furnished unspecified services. By the early 1920s, Cecelia LeBrun provided chemistry services and there were several laboratory assistants, some of whom served for only a year or two, while others had a longer attachment. As in Halifax, the evidence for most of these workers resides in the public accounts. Only rarely did staff changes merit mention in the annual reports to the health minister or in other documents.

Marilyn Clarke first arrived at the Bureau of Laboratories as a student and stayed for another four years as a member of staff. Like many of her Halifax counterparts, Clarke was educated at Dalhousie, receiving her BA in 1930. She was described as an "all-round girl" in her yearbook, excelling in her academic studies while "never miss[ing] a party." The yearbook also noted that Clarke wanted to "continue her research in Biology next year, a study in which she is intensely interested."[101] Clarke was a native of Saint John, the daughter of a local dry goods merchant. So perhaps familial connections led her to enroll in the laboratory course in that port city in 1930. Familial support was likely necessary: while training in Saint John, she was not paid.[102]

Clarke was familiar with most routine hospital work, including urinalysis, blood chemistry, hematology, bacteriology, and section cutting and staining. Her peak salary in Saint John was six hundred dollars a year.[103] Clarke left in September 1935, to assume a position at the Prince Edward Island Hospital in Charlottetown, where she remained until December 1937. Her replacement in Saint John was Stuart Johnson, who had been "apprenticing" in the facility for several years.[104] Johnson's salary shows wide variation. In 1932–33, the year of the first recorded salary, he earned a mere fifty dollars, which was increased to three hundred the following year and fell to $250 in 1934–35. In 1935–36, the year he replaced Clarke, Johnson's salary was increased to $550 for eleven months' work.[105] In all likelihood, Johnson was serving in the laboratory only part time, working irregularly and earning fifty dollars a month, the same amount paid to women in the laboratory with several years' experience.

In 1935, longtime employee Mary Peterson died. She had been considered "faithful and hard-working."[106] Peterson, a married woman, arrived in the laboratory in the summer of 1928, earning forty dollars a month, a figure that remained unchanged during her tenure at the bench.[107] There were a few other married women who worked in the laboratory.[108] Mrs Nellie Bush started in the laboratory in the midst of the depression, earning $41.33 for relief work. The following year she provided services while Mary Peterson was on vacation, earning $17.33. Subsequent years saw Bush earn small sums, though she assumed a regular appointment partway through the 1936–37 fiscal year and remained in service until 1945. Other married women worked only briefly in the laboratory. Iris Conrad, for example, left shortly after she married.[109] Still others, such as Emily Byrne, left laboratory work prior to their marriage. Byrne apparently worked on a peripatetic basis in the laboratory, if her wages are any indication. Beginning in 1931–32, she earned fifty dollars, while subsequent years saw her earn $292 and $536. By 1934–35, she was paid the full wage of $600 per year. Byrne was compensated for "laboratory services" according to the public accounts; she was never called a "technician" despite the "sterling" service she provided.[110]

Most of the women who laboured in the laboratory were single women. Two long-standing staff members in the laboratory were Muriel McCoy, who worked as a technician, and Phyllis Evanston, who was an office worker McCoy started in 1926–27 at an annual wage of $275, which increased in the closing years of the 1920s to $600. When she joined the Canadian Society of Laboratory Technologists (CSLT) in 1937, the seventh person from New Brunswick to do so,[111] her application noted that she was the chief chemistry technician, although she could also work in bacteriology or hematology, or prepare solutions for microscopic work. McCoy eventually achieved the title of "Chief Technician and Provincial Analyst" in the mid-1940s and, for many years, was a prominent member of the CSLT in the Maritimes. Evanston joined the staff in 1927–28 as a clerk and stenographer at a wage of $75 per month, which exceeded that paid to most of the other laboratory workers. This sum was raised to $1,020 the following year. There were modest reductions through the early 1930s to slightly less than one thousand dollars, although her wages were restored in 1934–35, and then increased to $1,100. Through the late 1920s and 1930s, Evanston was consistently the highest paid employee in the laboratory, after the director.[112] She kept the accounts for the laboratory, prepared the reports, and watched over the filing with only occasional assistance. These were all integral tasks for the work of the lab, insofar as accurate and timely reports were critical for clinical care, identifying and responding to areas of concern and compiling statistical data on specific diseases. By the early 1940s, there was too much work for one person and the provincial government made two additions to the office staff.[113]

That the one long-serving clerical person earned more than most of the bench workers is noteworthy. In Halifax and Saint John, employment opportunities for women were limited. Even elite women, such as the women doctors described earlier, faced significant obstacles in establishing their practices. Female physicians were channelled into "appropriate" spheres such as pediatrics or obstetrics. For many women laboratory work was respectable work, where those with an interest and aptitude for science could find employment. It did not enjoy the visibility of other kinds of women's work such as teaching or nursing, but it nevertheless appealed to many. It also offered reasonable salaries and security.

Yet laboratory workers were also vulnerable. Married women in Saint John earned less than either men or single women. The laboratories could always add students or recruit volunteers to perform laboratory tests, especially the routine analyses, a situation that undermined the security of paid workers. If women laboratory workers wanted to stay close to home, if they were indeed "local girls," they had few employment options. That Evanston earned more than her technical co-workers indicates the importance of clerical work for the laboratory's operation. The reports generated were numerous and had to be sent out quickly. As suggested earlier, reports were the productive unit of the laboratory. They communicated laboratory results to clinicians or public health departments and while bench workers would sometimes do this work, in the larger labs the volume of paperwork was substantial.

In smaller laboratories, the boundary between clerical and technical work was not as sharply demarcated as it was in Saint John. Nevertheless, in Canada both were for the most part women's work during the interwar period.

Technical workers were rarely men in the Maritimes, a pattern consistent with other North American settings. Ronald Burns joined the Saint John laboratory staff in 1929–30, before he celebrated his twentieth birthday. He was a graduate of Saint John High School and spent many years at the laboratory, eventually assuming the position of senior technician.[114] Burns later entered medicine and, by the late 1940s, he was a practicing physician, although he maintained his connections with laboratory work until his retirement from practice in 1977.[115] In the mid–1930s Burns was joined by Stuart Johnson, who also left to pursue a medical education.[116] While pursuing their medical educations, both men continued to provide much-needed relief to the laboratory during the war years.[117] But such men were merely sojourners through the laboratory, earning a few dollars to support their ambition. Overwhelmingly, laboratory work in health care was women's work.

"LABORATORY BOYS"

It would, however, be erroneous to conclude that all laboratory work was women's work. While women were staffing hospital and public health laboratories throughout the Maritimes during the 1920s and 1930s, young men predominated in the medical science laboratories at Dalhousie University, the other major employer of technical staff in Halifax. To be sure, the work in the two settings was different. Work in hospital and public health laboratories was predominantly oriented toward diagnosis, while university laboratories were primarily interested in education. Nevertheless, many of the tasks were similar and work conducted in one setting would be largely familiar to those labouring in the other. The historical evidence from Dalhousie therefore serves as an important reminder that similar work could be gendered in very different ways, even in the same city neighbourhood. Samuel Richards, a "laboratory boy" from north-end Halifax, began work in Dalhousie's physiology department in September 1925. He was hired on a trial basis for twenty-five dollars a month and proved "very satisfactory." In January, his wages were increased to thirty dollars.[118] Richards's work with the department ended on 1 May 1926 and Charlie Mitchum replaced him the following fall. Mitchum's monthly wage was thirty dollars and he was an exceptional addition, being both efficient and innovative. Professor Boris Babkin wanted to retain Mitchum for the following year and asked Dalhousie president A.S. MacKenzie if arrangements could be made to employ Mitchum through the summer. MacKenzie telephoned Babkin on 10 May telling him that the young man could be retained "if he could be kept busy." Mitchum would spend the summer assisting in the research work of Miss Betty MacNeil, taking electrocardiograms, filing, and maintaining animals with fistula. In September 1927 Mitchum's salary was increased to thirty-five dollars. Mitchum resigned in March 1929, although he agreed to stay on until the end of April, while a replacement was found.[119]

Dr E.W.H. Cruickshank tried recruiting one "boy" but when none materialized, he turned to seventeen-year-old Charles Livingston. Due to his tender age and "immaturity," Cruickshank suggested appointment on a probationary basis for fifteen dollars from May to July; if Livingston were found satisfactory, his pay would be increased to thirty-five dollars in August, "with assurance of [further] advance if worthy of it."[120] It did not work out and Robert John Dempsey was appointed to perform the work in July. Dempsey, however, decided to return to school in August and left at the end of the month.[121] There were others: Gregory James in 1929–30 and Daniel Mitchum, who was appointed in the autumn of 1930 and worked in the physiology laboratories until December 1931.[122]

Many of the laboratory workers in physiology were in their teens or early twenties, and their wages reflected this. But it was in the biochemistry department where the meaning of the wage structure found bold articulation. Biochemistry was in its formative period. The laboratory had five tables, which were able to accommodate six students on per side. There was very little equipment on hand, in part because until 1923, biochemistry, histology, and physiology were all one department, and the extant equipment was divided between the latter two. Shortly after receiving his appointment, Dr E. Gordon Young, who worked in biochemistry at the University of Western Ontario, wrote to MacKenzie to inquire whether there were laboratory technicians available in Halifax. MacKenzie suggested it "would depend so much upon the qualifications which *he* must possess."[123] In Young's estimation, there were two "types" of laboratory technician: "either a young man of about 20 or a middle aged man with moderate intelligence but without initiative. They can both be trained to be good technicians. I think that I prefer the former. A knowledge of elementary chemistry is desirable but very rare. It is however very essential that the applicant have a useful pair of hands and be naturally careful. In a few months work I can train him in ordinary manipulation."[124] An examination of two of the young men employed in biochemistry, Stephen Brown and John Grey, reveals the limitations and frustrations of laboratory work in the medical sciences and male workers trying to earn a living.

Stephen Brown, a young man of about twenty, began to work in biochemistry in October 1925 at a salary of twenty-five dollars a month. By early 1926, he had proved himself "intelligent, most reliable and teachable" and Young suggested an increase to thirty dollars. Despite another raise in September 1926 to thirty-five dollars, Young believed that although Brown enjoyed his work, he was "discouraged." Young told President MacKenzie that Brown was threatening to leave "unless he can see a living wage and some prospects ahead of him." Now twenty-one, Brown believed his age and experience warranted a wage of forty-five or fifty dollars a month.

Equally important for Brown was a promise of regular advancement, which suggests that, for this young man, laboratory work might have been more than a temporary job during the passage to adulthood. His aspirations enjoyed the support of Young, who appreciated his services. Young suggested in early 1927 that without Brown, there would either be no time for research or a "trained faculty assistant" would be required, presumably a more expensive option. Indeed, it was only Brown's extraordinary capabilities that permitted the absence of such an assistant. MacKenzie raised Brown's wages in March to forty-five dollars per month, to be increased to fifty dollars when he celebrated his second anniversary. With continued progress through "his industry and improved skill" he could expect sixty dollars if he was deemed to be "worthy of being retained and advanced."

Brown achieved this wage in October 1928, his fourth year in the biochemistry laboratory. His effort was extraordinary. He often worked on Saturday afternoons and Sunday mornings because of the large class sizes. He also trained himself to blow glass and work metal, thereby saving the university money for new apparatus or employing craftsmen. Perhaps because of his obvious initiative and a sense of frustration over his paltry wages, Brown was growing restless. Young reported that he felt he should "be getting a living wage at least comparable with the boys of his own age" and that "it would be a severe loss if he were to leave." Brown resigned in June 1929.[125]

He was replaced by eighteen-year old John Grey, a grade ten graduate who was appointed in September 1929 at thirty-five dollars a month, and he received an increment of five dollars the following month. Young reported that, like Brown before him, Grey "likes the work and is learning fast." Young apparently learned from Brown's resignation and he suggested that "from the standpoint of the University it is very essential that he should be contented financially, as good boys are hard to get and hard to train," while noting that there would be a need to consider further salary increases. President MacKenzie concurred and increased the monthly salary to forty-five dollars in May

1930. MacKenzie suggested that "laboratory boys" needed something to "stimulate them to do better work" and suggested that if his work continued to be satisfactory, his salary would increase to fifty dollars.

MacKenzie and Young agreed that the university would never pay a wage "suitable" for a married man, even though they recognized that staff turnover resulted in lost productivity. MacKenzie explained to Young that the university "cannot afford to pay a salary to a laboratory assistant that will keep a married man." When this was explained to Grey, he was "very discouraged." His disappointment must have been heightened, for he was in the midst of preparing for marriage. But Young's intervention carried the day, and MacKenzie was persuaded to increase Grey's salary by another five dollars, effective 1 April 1931, "with the expectation that he will look and find a job" that would provide for him (and presumably his future wife and family). He charged Young to make it clear to Grey that there would be no further increases, and that Grey "should begin at once to look for some more remunerative work."[126]

Despite his January 1931 marriage, Grey did not leave the biochemistry department at the end of the 1930–31 academic year, as MacKenzie suggested. In correspondence with new Dalhousie president Carleton W. Stanley, Young argued that "I am entirely opposed to the policy of starving our laboratory assistants so that it is necessary to train new ones every two or three years with a decreased efficiency in the preparation for classes and a decreased research output on the part of the professor. All this is for a saving of a few hundred dollars a year."[127] Stanley promised to raise the matter at the next executive meeting of the board, and while there is no recorded response, a budget dated 29 December 1932 shows that $780, or sixty-five dollars a month, was to be expended for a laboratory assistant during 1933–34. Young managed to retain Grey through the rest of the 1930s for seventy-five dollars, the maximum the university would pay for technical workers.[128]

The debate surrounding the wages for the marriage-minded John Grey did not prevent the appointment of another married man.[129] In December 1931 the physiology department hired Albert Hallett, who had five children. Hallett had "considerable training in hospital technique" and came well recommended. Because of his age, family situation, and experience, Professor E.W.H. Cruickshank requested that he start at fifty dollars a month. Stanley turned this down, noting that Hallet's predecessor had earned only forty-five dollars a month, which was all that had been budgeted.[130] When Cruickshank renewed his request nearly a year later, he reminded the administration that Hallett was married and had a family to support, adding that only the eldest boy was contributing a wage to the maintenance of the family.[131]

The medical science departments were content to employ you to perform the necessary tasks for the university to conduct [research] and teaching. The medical school of the 1930s simply co[uld not] function without workers such as Brown, Grey, Hallett, or countless other technicians or assistants in anatomy, biochemistry, pathology, and the other medical sciences.[132] The wages paid to Dalhousie's laboratory workers reflected both their young age and the perception that they were readily replaceable. The only requirement was that they possess a good pair of hands and be naturally careful, because training could take place at the bench over several weeks. The university's reluctance to hire married men supporting families reinforced the idea that laboratory work at Dalhousie was transitory. Demands for wage increments, cast in the language of a family wage, were often rebuffed and, as a consequence, male workers moved through the medical science laboratories with some regularity.

The employment of men in the Dalhousie laboratories, which was deeply inscribed by gender, is distinctive and instructive. Employees in the Dalhousie laboratories were usually in their late teens or early twenties. The paltry wages and poor chance for advancement discouraged some of the workers and there were frequent staff changes. Yet staff turnover and the loss of productivity did not emerge as significant issues for the university, even with the apparent difficulty professors such as Gordon Young had in finding competent personnel. Research in the university, which would have demanded some stability in the staff, was not yet pre-eminent at Dalhousie. While Young may have desired consistency, the administration did not view expenditures for technical hands as a priority. What the professors and the administration agreed upon was that Dalhousie would never pay a wage sufficient to support a married man. Ambitious men, according to Young, would not be attracted to the work. Only young men or those without initiative would submit to the low wages offered. The gendering of laboratory workers at Dalhousie was underpinned by the idea that these workers were not men supporting families, despite the fact that some already supported children or aspired to married life.

CONCLUSION

The creation of a labour supply for the new service departments that came to define the hospital through the 1920s, 1930s, and 1940s was guided by a variety of factors. Physicians directed these services but trained staff carried out routine tests. At the same time, physicians positioned themselves as the *interpreters* of diagnostic results, a point to which we will return. This separation of the mental and manual aspects was deepened by the treatment of women who became, to use Stevens's phrase, "all-purpose service workers." Nurses and other women were to be "all round" employees, capable of doing any task within the hospital. Although there was a clear emphasis on fulfilling a variety of roles within the growing hospital complex, there was also a growing demand for formal courses of instruction. The competing imperatives of multitasking and specialization coexisted in the laboratory and, by extension, throughout the hospital.

Another element present in the discussion of the Dalhousie laboratories is what David Montgomery has called the "historic discovery of labour turnover."[133] Both Gordon Young of the medical school and A.G. Nicholls of the Pathological Institute suggested at different times that staff turnover limited the productivity of their respective laboratories. The front-end savings realized through inadequate wages was countered by the cost of continually having to train replacement workers. That is to say that cheaper workers were expensive to replace. Without a defined skill-set or standardized education or apprenticeship, responsibility for training replacement workers fell to senior staff members or laboratory directors, with an attendant loss in efficiency, productivity, or research activity. In contrast, highly skilled workers (such as those with either significant education or experience) could be replaced because their skills were well honed and portable, hence there was a shorter learning curve. The management strategy thus became one of paying a wage sufficient to maintain the staff, but recognizing that others could assume the bench work if wage demands became excessive. Nationally, the concern with retaining staff gave rise to innovations in management, such as personnel departments and corporate welfarism. Locally, departures were common as employees found more remunerative work, or retired in favour of other pursuits. Concern for staff turnover was a short-lived manifestation of labour scarcity. By the later 1920s and through the 1930s, laboratory workers also had to contend with the annual presence of volunteers or students. As in hospital schools of nursing these students could be an important source of labour, easily moulded to respond to increasing workloads or new innovations.

The debate over wages for the employees of the medical school signifies the important difference between some male laboratory workers at Dalhousie and their female counterparts at the bench. Clearly, men such as Stephen Brown and Albert Hallett were articulating a need for a wage sufficient to maintain their families. As Joan Sangster has suggested, the origins of the family wage ideal are disputed, though the consequences were clear, particularly for women. The notion of a family wage "constructed an image of women as dependent and transitory workers, thus making them more dependent; and it ignored t[...] ties of women who were self-supporting or were the sole su[...] their families, or whose husbands were unemployed, tempor[...] permanently."¹³⁴ In Saint John, the married women who worked in the laboratory earned forty dollars a month while their single co-workers, both men and women, earned fifty dollars a month. Beyond gender, the other significant factor in determining wages for laboratory workers concerns status. Workers were attempting to claim that they were, in fact, highly competent and skilled workers. The CSLT's search for professional status, explored in the next chapter, was predicated on this view of laboratory work. However, laboratory workers' claims to authority were consistently undermined through the employment of volunteers, summer students, or others who lacked training. More critically, the structure of laboratory work, which left interpretation of results to physicians, enshrined their role as helpers, subordinate to organized medicine.

Finally, opportunities for women were very much constrained. There were few options for women in any profession. As the brief discussion of female physicians who graduated from Dalhousie illustrated, even highly educated women faced significant career obstacles. While many of the laboratory workers enjoyed a good education, the prospects for a career remained limited. For many, a job in a laboratory was a welcome alternative to caring for their families, tending the sick on the ward, hawking goods in department stores, or teaching. It was also a job that provided an outlet for women with an interest in and aptitude for technical work within science. Women used a variety of strategies to identify opportunities at the bench. In Halifax many workers had a Dalhousie connection, with some holding degrees that brought them into contact with the laboratory directors. Other workers used a wide variety of means to secure positions, including personal connections and neighbourhood information about marriages and job openings. For these women, the laboratory was not entirely a dead-end career. It provided a choice for women interested in science, albeit in a limited fashion.

5

Diffuse Roles and Multitasking

"Oh she's a gal who fares far worse
Than any stiff-starched graduate nurse."[1]

Laboratory workers in Saint John or Halifax, and those who staffed the benches in rural hospitals around the Maritimes, came from a variety of backgrounds. Some were nurses, others held university degrees, while still others received training in the laboratory following secondary schooling. All laboratory workers learned new procedures at the bench and occasionally they were sent elsewhere to become skilled in the latest techniques. Workers from rural hospitals would travel to Saint John or Halifax, while workers from those cities would occasionally go to Toronto or American centres. There was a baffling array of expertise and personnel within any one laboratory. A worker at the federally operated Lancaster Hospital in New Brunswick wrote that there were "nine technicians, three laboratory helpers, three students, a biochemist and a research assistant" by the end of the 1940s.[2] Some laboratory workers took responsibility for individual sections. In the Bureau of Laboratories, for example, Rachel Hunter and Dorothy Jakeman worked in serology, Cathy Arnold was responsible for histological work, Evelyn Russell for chemistry, and Jean Hayes for hematology, while Margaret Bryden and Dorothy Tapley shared the heavy workload in bacteriology.[3] The experience of these individuals varied. Hunter had worked for several years by the early 1940s while her coworker in the bacteriology section, Dorothy Jakeman, was a recent appointment. The workers were variously described as technicians, assistants, or employers who were paid for "services." The exception to this was Bryden, who was described in the public accounts as a bacteriologist, as was her predecessor, Patricia Carew.

It is exceedingly difficult to interpret the meaning of these designations or to understand completely what work individuals performed and the relationship among different aspects of laboratory work. In Halifax, there is little question that Margaret Low, during her long tenure at the Pathological Institute, was the senior technician. She exercised supervision over much of the daily work and, when the director was absent, administered the laboratory on his behalf. In community hospitals the staff worked independently or with limited supervision. For many laboratory workers, the nature of their work is difficult to delineate concretely. "Lab services" or "lab assistant" are decidedly ambiguous terms. Clearly, they encompassed the execution of laboratory analyses. But in 1944 when the Saint John laboratory appointed Mae Bell in the glassware department, her work was deemed to be "laboratory services" as well, while the next year she was acknowledged as a laboratory assistant.[4] Was she performing tests or cleaning glassware for the laboratory? The historical record is not sufficiently complete to draw a conclusion. What is clear is that many workers had very broad responsibilities and diffuse roles.

CROSSING OCCUPATIONAL BOUNDARIES

Isabel Robinson presents an interesting case both because of her peripatetic journey through the Halifax laboratory and the work she performed. She held "temporary appointments" for many years and provided "valuable services" preparing diagnostic outfits and as a relief stenographer.[5] Robinson's presence reminds us not to draw too sharp a distinction between clerical and technical workers in these early decades. Further cautionary evidence comes from Deborah Henderson, who worked at the Victoria General Hospital as a "Stenographer Path Tech" in September 1925. Payroll records note Henderson's role as a stenographer on 15 July 1927, but in the very next payroll period, ending 31 July, she was designated a pathological technician.[6] Later in 1927 she was described as a "part time technician" at the laboratory.[7] Such descriptions appear to be largely arbitrary and therefore suspect. In another example, Dorothy Jakeman's first foray into the Saint John laboratory was in 1939–40, when she earned slightly over forty-five dollars for relieving the caretaker during his illness. The following year, she joined the staff as "lab technician."[8] Changing roles, or moving between them, was typical for laboratory workers during the first half of the twentieth century.

The examples of Robinson, Henderson, and Jakeman demonstrate the fluid rhythm of laboratory work and identities. When the public health work for Nova Scotia was hived off from the pathological section in 1926, the hospital laboratory workers lost their stenographer. Typing duties subsequently fell to the existing staff. Margaret Rogers joined the staff directly from high school and she was therefore several years younger than her co-workers. She was assigned responsibility for typing all the pathological reports. When Rogers was absent, as she once was with a lengthy middle-ear infection, clerical duties fell to other workers. Sixty years later, Ellen Robinson recalled, "I could *still* type those [reports] ... I used to start with the gross and histological appearances of such-and-such."[9] While short, many reports had to be sent out the same day that a specimen was received. The head technician of the Cape Breton Public Health Laboratory described the extent of these duties to a colleague in 1950: "Well, among other jobs here I am doing all the typing of reports. I typed 23 letters alone before 9 a.m. the other morning. It is just 7.30 a.m. now. This getting up at 5.45 am. isn't so hot, believe me ... I have a lot of reports to type so they can catch the mail going out at 10 so I'd better hike."[10] Reports were the productive unit of the laboratory, the essential link between the science of the laboratory and the clinical diagnosis or management of the case. The onerous production of these essential reports, which were as critical as the actual laboratory analyses insofar as they communicated information to attending physicians, fell to the workers in what was a familiar pattern of multitasking.

The women who laboured at the bench moved easily among various tasks, including different kinds of laboratory analyses, clerical jobs, or cleaning glassware. Typing and record keeping were part of the work for most laboratory workers, but occasionally the boundaries between entire hospital departments were blurred and workers found themselves with wide-ranging duties. No less an authority than *Canadian Hospital* suggested in 1930 that in smaller hospitals, pharmacists might also perform selected laboratory tests or serve as the x-ray technician for the hospital.[11] In 1920 one male hospital workers' day was organized to dispense drugs in the morning, then develop x-ray films for an hour, while reserving the afternoons for laboratory work.[12] Many workers had similar days, filled with an assortment of responsibilities. In 1930 the Mirimachi Hospital in Newcastle purchased new x-ray equipment and requested one of its nurses, who had graduated earlier in the year, to take a combined course in x-ray and laboratory training.[13] Alice Thorngate, in a rare published account of laboratory work, offers a clear illustration of the variety of roles laboratory workers filled in Wisconsin:

Times were rough in the 1930s and laboratory jobs were especially hard to come by. I tried various kinds of work while waiting for something in the laboratory field to open up ... Eventually I found a position as technician and general office girl in a doctor's office in central Wisconsin. Beside doing blood counts and examining urine specimens, I took simple x-rays, acted as receptionist, kept the books, took care of correspondence, made out insurance forms and did surgical dressings on fingers and toes.[14]

When the Chipman Memorial Hospital in New Brunswick appointed an anesthetist in 1922, they sent one of their own nursing graduates, Margie Fitzpatrick, to Lakeside Hospital in Cleveland, Ohio, to pursue training. Fitzpatrick, who had been working in Arizona as a nurse, returned to Chipman in October following receipt of her anesthetist diploma. When the laboratory equipment was upgraded at a cost of several hundred dollars the same year, responsibility for that work also fell to Fitzpatrick. In the two years that followed her graduation in January 1921, Fitzpatrick had gone to Arizona to nurse, Cleveland to further her training, and back to Chipman, where she assumed responsibility for both laboratory and anesthesia work.[15] The path from training to appointment as a nurse to work in other services was well worn during the 1920s, 1930s, and into the 1940s.

Irene Mellish was born in Halifax in 1909, the eldest of six daughters. Her father had wanted to be a physician but worked instead for the local Bible Society. Mellish pursued nurses' training at the Victoria General hospital, beginning in September 1929, after completing grade ten. Mellish later became the superintendent of Eastern Kings Memorial Hospital in 1942, where she remained for twenty-five years. There she "was responsible for management of the whole hospital, right from the top right down to the bottom, lowly janitor ... we didn't have a pharmacist, we didn't have a dietitian ... [We had] nurses and cleaning staff."[16] With such a limited staff in community hospitals, it is not surprising that workers fulfilled multiple roles in a variety of services.

Mary Kathleen Murphy was born on 1 February 1906 in Sydney, the daughter of a steelworker and a dressmaker. She obtained her grade eleven and began her nurses' training at the New Waterford General Hospital in September 1925. It is interesting that in recounting her work, she was asked whether or not she cooked. Her response was, "Ourselves? No, we had a cook and an intern and a maintenance man. We had everything they have today, actually." But clearly they did not. For example, x-rays and laboratory work were tasks for nurses in New Waterford's hospital. The prevailing pattern of work for nurses included such duties. Murphy recalled, "We had a Miss A.J. MacDonald, she was next to the Superintendent and she was in the laboratory, she *of course* took x-rays and she did all the work like urines and sputums and she was really something. It was funny, she took me and used to give me a lot of training in the lab. And I'll never forget what she taught me, she taught me to take x-rays." Murphy, despite her nurse's training, spent two decades "taking pictures."[17] Yet in recalling her work experience, she did not consider this unusual or unexpected. As Margarete Sandelowski has recently argued, "Nurses did whatever was necessary and what others did not want to do."[18] That nurses would fill roles in other departments, even for years, suggests that the demand for labour quickly consumed or challenged any exclusive professional identity in many hospitals.

Indeed, nurses' familiarity with various hospital departments began during their training. Nursing schools in the Maritimes, as elsewhere, were administered by hospitals and the education of students was often secondary to the demands for workers in the hospital wards.[19] Sister Catherine DeRicci, who entered the Halifax Infirmary nursing school in 1927, recalled that senior students would often be found in the operating room, x-ray department, and "all those places,"[20] presumably including the laboratory. As Susan Reverby suggested, "nursing education was called training; in reality it was work."[21] In Halifax lectures were scheduled around twelve- or thirteen-hour shifts. Students often missed lectures because of ward duties and the lecture schedule varied both in quantity and quality, depending upon the availability of doctors.[22] Indeed, the workload of apprentice nurses had become so substantial that in his influential 1932 report George Weir felt compelled to comment that nursing schools "should be considered primarily as an educational institution rather than as an economic asset to the hospital."[23] In the 1920s and 1930s, students were clearly an integral component of the expanding hospital labour force. Nurses had to discharge their duties on the ward floors, perform a variety of domestic tasks, complete administrative duties, and, increasingly, work in other patient services, including the laboratory. Throughout the 1920s, 1930s, and 1940s, hospital workers, including nurses (arguably the most specialized and recognized of the emerging health care professionals), performed a broad range of duties.

MANY PATHS TO THE CSLT

The registry information of the Canadian Society of Laboratory Technologists confirms and extends the experience of nurses.[24] By the 1930s, women had gained the right to practise in all the professions, though their presence was minuscule in most fields. Several professions even utilized quotas to limit the number of women in their ranks.[25] Observing the professionalizing efforts all around them, a dedicated cadre in Hamilton, Ontario, embarked on a project to organize laboratory workers. On 8 November 1936, eight interested individuals met in the Hamilton General Hospital and founded the CSLT.[26] Among the society's six stated objectives were improving the "qualifications and standing" of laboratory workers and, once this was achieved, promoting "a recognised professional status" for the workers. The founding members believed that establishing approved training facilities and

Figure 5.1
Growth of the CSLT, 1937–1950

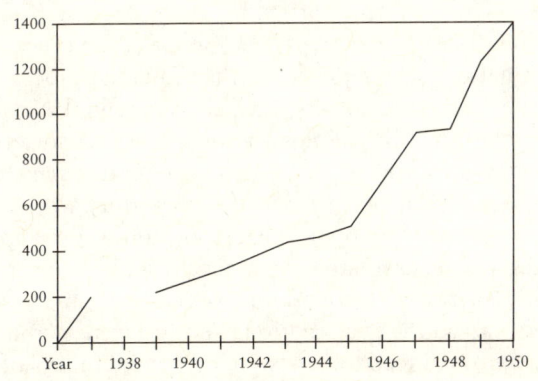

Source: CSLT. Minutes of Annual General Meetings 1937–1950. No data were available for 1938.

national practical and theoretical examinations would provide a vehicle for professional uplift.[27] The imposition of exams, registration, and training schools are among the key hallmarks of professional identity in health care and beyond. Yet what is most intriguing about the history of laboratory workers is that while the professional project was being vigorously pursued, there were many paths to working at the laboratory bench and to membership in the CSLT. Sociologist Anselm Strauss challenged the idea of homogeneity within professions forty years ago, when he wrote that "there are many identities, many values, and many interests" within any one occupational group.[28] Certainly, laboratory workers bear this out. Laboratory workers had a diffuse identity within the laboratory and hospital, fulfilling a wide array of tasks and staffing numerous services, and these various identities and values were reflected in the CSLT's membership.

The lack of homogeneity identified by Strauss is readily apparent among the CSLT's charter members, which included seven bench workers, an analytical chemist, a surgeon, a physician, a pathologist, and a secretary.[29] From this small beginning, the CSLT began to grow almost immediately and by February 1937 it had twenty-three members. It was also "national" in scope. By the first annual general meeting, held in a dining-room in Hamilton's Royal Connaught Hotel and attended by 27 keen individuals, there were 193 members drawn from every province. Half of the membership came from Ontario.[30] A decade later, the CSLT could boast twelve hundred members from across Canada with provincial branches in British Columbia, Saskatchewan, and New Brunswick, although Ontario continued to dominate membership (See figure 5.1).

The growth of the society, while impressive, masks a complex history for the rank-and-file worker. While circulars dispatched from Hamilton extolled the virtues of the national society and invited laboratory workers to join, many were sceptical. "What will be the advantage of taking examinations for registry in the Canadian society or of belonging to the Canadian society," asked a potential New Brunswick member, "if one already belongs to the American society[?]"[31] The American Society of Clinical Pathologists (ASCP) first registered laboratory workers in 1928. In 1933 American laboratory workers organized a national society of their own, but responsibility for approving programs and registering graduates remained with the ASCP.[32]

The ASCP also routinely registered Canadians. In 1936 two Saint John workers passed the examinations and received accreditation from the ASCP as "medical technologists." The annual report of the Bureau of Laboratories suggested that achieving the "MT" designation was "becoming increasingly important and will serve in time to place trained laboratory technicians in a more secure position in competition with those whose training has been superficial."[33] The next year, two more workers received their MT, bringing to four the total registered with the ASCP.[34] Even after the creation of the CSLT, several workers in Saint John wrote the American exams and registered with the ASCP.[35] The annual report for Nova Scotia does not record similar instances, although one informant did recall that three workers in the laboratory sat for the American exams.[36] Nevertheless, registration was unusual for early laboratory workers. Among the earliest thirty members of the CSLT from the Maritimes, all of whom joined in 1936 or 1937, only three held ASCP memberships.[37]

Those who advocated the creation of a national society had a tall order in convincing others of the merits of membership. Laboratory directors who trained their own staff did not welcome the intrusion, nor were laboratory workers convinced of the value of a national society. Moreover, those who worked in laboratories were hardly a uniform lot. Despite all these caveats many did seek membership in the Canadian body after its creation. Of the first two hundred individuals, all of whom joined in late 1936 through 1937 and the first months of 1938, 158 (seventy-nine percent) were women. The early members had a wide variety of educational backgrounds and their duties were similarly diverse. Some worked in small rural hospitals where only a few simple tests were conducted, while others joined from large urban general hospitals with highly specialized services. Some of the early members laboured exclusively in the laboratory, while many combined their work at the bench with duties throughout the hospital.

A senior member of the Charlottetown Hospital, who joined the CSLT in 1939, provides an illustration. Her application form notes that she supervised not just the laboratory but also the medical records and pharmacy departments. She was well prepared for all these duties, holding certifications as a record librarian and a medical technologist in addition to a pharmacy degree. As she assumed new responsibilities within the Charlottetown Hospital, this individual pursued both educational opportunities and professional credentials. Many other registrants similarly combined work in different services. A woman at Dawson Memorial Hospital in Bridgewater, Nova Scotia, performed laboratory and x-ray work. Her laboratory duties were restricted to general work in bacteriology, hematology, and urinalysis tests, while other analyses were done in Halifax. A Catholic sister in charge of the laboratory at St Mary's Hospital in Inverness, Nova Scotia, also took charge of the x-ray department.

The common expectation that hospital workers would work across departmental boundaries was of concern to some individuals. A nurse from Colchester County Hospital in Truro, Nova Scotia, was responsible for routine laboratory tests, including blood counts, fecal examinations, urinalyses, sputum, blood groupings, and a variety of other tests. Other analyses, such as all the pathology work, serological tests for syphilis, and some blood work, were sent to Halifax. In addition to the laboratory work, the nurse also served as the x-ray technician and, in her capacity as nurse, was in charge of the operating room. Despite these broad responsibilities, the nurse was by no means sure of her abilities in all these areas. After eighteen months in the CSLT, she wrote to the national secretary, Denys Lock, asking whether there was a one-year course in laboratory technique "for people who already have some knowledge of the subject, but feel that their training and experience is not sufficient."[38] In the spring of 1939, there was no such course available in eastern Canada.

The career of individual workers illustrates only the range of duties waiting to be filled within the modernizing hospital. It says nothing of the meaning of these opportunities, either for the individual or for entire occupational groups. The example of nurses who worked in laboratories, because they enjoyed a degree of professional recognition (unlike laboratory workers who were still in the midst of their professional project), permits some interpretation of the effect of multitasking. In her summary of the wide-ranging literature pertaining to nurses and medical technology – including work by Barbara Koenig, Anselm Strauss, and David Wagner – Margarete Sandelowski notes that medical technology has "been charged with reinforcing the subordination of nursing to medicine, impeding its development as a valued province of knowledge and practice, and undermining its very essence."[39]

Nurses' responsibilities in x-ray or laboratory departments, duties often shared with non-nurses, undermined the distinctiveness of nursing. Conversely, the addition of diagnostic services to hospitals, and other departments such as medical records, pharmacy, dietetics, and occupational and physical therapy, contributed a great deal to the professional enhancement of medicine, both scientifically and materially. In many hospitals, however, the actual work in these departments was combined with work elsewhere. In other words, the multitasking of women facilitated the modernization of hospitals throughout Canada, but it did little to enhance the status of the women themselves. In fact, the opposite is true; many hospital workers shared responsibility for departments with others who may or may not have had similar education or training. The end result was that their activities could not be clearly distinguished from the work of others in the hospital. To a degree, they became interchangeable, though all were clearly subordinated to professional medicine in the hierarchy of health care.

The evidence clearly demonstrates that individual workers had duties in the hospital that transcended occupational boundaries. Hospitals wanted workers who could meet their expanding needs, and do so cheaply and efficiently. Whether or not a hospital was large or small, serviced a rural or urban area, or was a general or specialty hospital determined what kinds of workers were needed and desired. But evidence of multitasking among women hospital workers is plentiful and national in scope. In 1930 the Aznoe employment agency advertised in *Canadian Hospital* for a "Laboratory x-Ray Technician, able to do blood chemistry" for a position in eastern Canada.[40] Prospective employees frequently marketed their range of skills. An advertisement for a "Woman Laboratorian" emphasized eight years of excellent experience, in addition to her nurses' training. A thirty-two-year-old "Nurse-Laboratorian" highlighted that she had "trained under outstanding pathologists" and had nine years' experience. Another "Woman Laboratorian" had trained for one year under a pathology professor and had ten years' experience.[41] A resident of Drumheller, Alberta, emphasized that she was a registered nurse but also "qualified" as a laboratory technician. In addition, she offered a prospective employer training in x-ray and cardiographic work.[42] A hospital in Lethbridge, Alberta, searched for a "nurse with laboratory training and experience."[43] An advertisement from the Colchester County Hospital in Truro, Nova Scotia, sought an "x-Ray and Laboratory Technician," adding that "a graduate nurse is preferred."[44]

Some individuals, such as Mildred Dobson of Winnipeg, emphasized their college education, in her case chemistry, and their technical abilities. Dobson was not only "capable of taking full charge" of the laboratory but also offered experience in routine laboratory analyses, Widals, bacteriological work, blood chemistry, and basal metabolisms.[45] The few men searching for positions used similar strategies. J.G. Truax of Hamilton, for example, emphasized his training at the Northwest Institute of Medical Technology in Minneapolis, in "laboratory, x-Ray, physiotherapy, and basal metabolism technique."[46]

The empirical evidence from the CSLT membership files, advertisements in national health journals, and the work experience of the Maritime women reveal an enduring pattern of working across hospital departments. In contradistinction to most hospital histories in Canada, which detail a process of increased departmentalization and specialization through the twentieth century, the example of laboratory work demonstrates that the boundaries were more fluid than many accounts would lead us to believe. A *Canadian Hospital* item on an amendment to the Nova Scotia Pharmacy Act commented in passing that pharmacists in smaller hospitals who "had the necessary training" could find themselves performing routine laboratory analyses or serving as the x-ray technician. Such pharmacists, assumed to be women, would in this way demonstrate "a wide field of usefulness to the medical staff."[47] Another commentator noted that smaller hospitals combined the work of the dietitian with "those of the housekeeper, the laboratory technician or the laundry supervisor," although she charitably conceded that in such cases "it is quite possible to overwork the individual." Another hospital, with an average patient census of 134, combined dietetic and lab work, "but only because the dietitian is personally responsible for the blood counts alone."[48] A historiography that emphasizes an unfettered, linear process of specialization obscures the complexity of hospital work. Moreover, it is a portrayal rooted in assumptions based on class and gender, largely informed through the examination of hospital-based physicians.

There is abundant evidence, drawn from many local and national records, that the notion of the laboratory worker remained diffuse. Michael Katz has suggested that when positions are added to an institution over time, confusion over the definition of roles arises.[49] As a result, how duties are defined becomes obscured and this overlap creates tension within groups and among them. In some hospitals, laboratory work grew slowly, from a few simple urinalyses to a larger variety and volume of tests. Others, such as the Bureau of Laboratories or the Pathological Institute, were carefully planned facilities, designed to provide services for New Brunswick and Nova Scotia respectively. Laboratory work was a planned extension of the modern hospital. It simply could not exist without the purchase of equipment, the preparation of reagents, and adequate staff. To function, laboratories needed the cooperation of departments of health, clinical departments, and physicians. In other words, the development of laboratory service, even a small one, required considerable planning. Still, the workers in the service performed a wide variety of roles.

The labour process in hospitals in the early part of the twentieth century was shaped, according to Susan Reverby, by two twin concerns. First, hospitals had a desire to maintain a stable workforce. This did not always translate into high wages for the workers, however. As discussed earlier, wages were determined through a careful consideration of the local economy, gender, and the cost of replacing workers. Reverby also suggests that the labour process in hospitals was shaped by organized nursing's desire "to establish its professional status."[50] Hospitals had long utilized an informal division of labour that saw workers in one service assume duties in another, as the situation demanded. As the workload in these services increased, or as beds were added to the hospital, the division was extended and then formalized, and new categories of workers began to emerge.[51] Before the Second World War, in response to changing definitions of what constituted patient care, nurses assumed a wide variety of tasks.[52]

It was not until the postwar period that registered nurses were joined by ward aides, licensed practical nurses, and registered nursing assistants. These "subsidiary workers" performed tasks formerly done by graduate nurses and those still enrolled in nursing schools.[53] Junior nurses regularly performed cleaning duties before World War II, while after the war such tasks became the responsibility of a completely separate service "and are now not considered nursing tasks at all."[54] Concurrently, nurses themselves began to assume new responsibilities and duties that facilitated a heightened specialization. A wide variety of courses were offered in such areas as x-ray technique, obstetrical nursing, or operating-room technique, and employees were offered bursaries to attend the courses, which lasted from four months to two years.[55]

As new tasks were assumed and old ones relinquished by various occupations within the hospital, the boundaries between groups became a matter for discussion.[56] How the content of the work, which had implications for staffing levels, worker satisfaction, and opportunities for advancement, was negotiated became a key feature of the modern health care complex. In the twentieth century, this has led to competition and conflict among groups of health care workers who are predominantly female.[57] Sarah Jane Growe wrote at the beginning of the 1990s of nursing's changing relationship with technology and with other health care occupations in a chapter tellingly entitled "Turf Wars." The use of more technology more often by nurses transformed the "content and design of nursing practice," as the distance between nurses and patients widened. Another transition was also afoot. Nurses in the postwar period increasingly had to work with other health care workers, such as occupational or physical therapists, speech language therapists, respiratory technologists, and a range of others. According to Growe, such patterns of care posed new challenges to "the already ill-defined boundaries of nursing work."[58]

Individual nurses responded differently to the multiple demands placed upon them. One author, writing in *Canadian Nurse* in 1941, commented that small hospitals might enjoy an adequate staffing complement for nursing but could very well lack people in the laboratory and other services. Work in those departments, together with record keeping, inevitably fell to the nurse.[59] As the oral evidence from Nova Scotia suggested, it is not at all clear that nurses considered these duties to be outside the field of nursing. Grace Cann returned to nursing in 1946 near her hometown of Overton, Nova Scotia. She had not nursed for five years and unexpectedly assumed the position of second night supervisor following the death of her husband. Cann found the night work a strain and had difficulty sleeping during the day, with the noises of downtown Yarmouth filling her room. She switched to day work, rotating through all the wards before finally landing in the laboratory. Cann worked there exclusively for over a year, before assuming responsibility for the children's ward. Interestingly, while she "took charge" of the ward, Cann also "helped" with medical records, "did some work in the lab," and dispensed drugs.[60]

A nurse writing in *Canadian Nurse* considered work as a laboratory technician, in the x-ray department, or as a record librarian to be "good fields" for the graduate nurse. In laboratory work, women could occasionally escape the demands of patients and physicians alike to a large extent, a freedom that undoubtedly appealed to some of the women. The author ended by noting that nurses were "being shut out more and more from choice positions" and "losing out in the hospital because she is not willing to prove that she can do better work than those who are not nurses."[61] Sister Catherine Gerard made a slightly different observation from her perch at the Halifax Infirmary. Writing in 1948, she lamented the declining quality of nursing service, placing the blame on the expanding opportunity for nurses in government employment and "in those hospital departments which formerly did not require nurses – for example, the x-ray department and laboratories."[62] Gerard believed that such opportunities were draining potential staffers at a time when there was a nursing shortage in Halifax. Gerard's comments also suggest that the lines of demarcation between work within the hospital were growing increasingly rigid as the many women who formerly pursued these tasks in addition to their nursing duties were simply not nursing at all, opting instead to work exclusively in other fields within the hospital.

CSLT MEMBERSHIP

In August 1942 a twenty-year-old Moncton woman began her training at the Bureau of Laboratories in Saint John. Like the students of the 1920s and 1930s in Saint John, Halifax, and elsewhere in the Maritimes, Muriel rotated through the various sections. Three months in hematology, two months each in bacteriology, serology, and histology, a little longer in biochemistry, while she spent several weeks learning the techniques of urinalysis, parasitology, and media and stains. In October 1943, after fourteen months of training, Muriel sat for and passed the Canadian Society of Laboratory Technologists registration exam. There was nothing remarkable about her training or experience. Not until she attempted to register with the CSLT did she encounter a problem. The Moncton woman had not completed her senior matriculation.

She was not alone.[63] Another woman training in Saint John had, like Muriel, failed to attain her grade twelve certificate. Helen was also refused membership in the CSLT despite having completed twenty-two months of nurses training.[64] At thirty years of age, Helen had already worked in the Bureau of Laboratories on two previous occasions, in November 1937 and January 1938, and spent one year working in a doctor's office. The director of the Saint John laboratory, Dr Arnold Branch, suggested that Helen was "very good in practice" and he had "no hesitation in recommending this young woman both to her ability and character." The applicant had already secured a position in a hospital, thanks to her experience, education as a nurse, and abilities.[65]

According to the admission rules set out by the CSLT and the CMA, students were supposed to have completed high school education before embarking on laboratory training and certainly before they were entered into the registry. That they were admitted to training and had work experience suggests that the rules were not rigidly adhered to in many settings. The experience of Helen or Muriel also reveals the limitations of such formal criteria as admission standards or training programs. Muriel had, after all, successfully completed both the training course and the national exam. She had also completed three and a half courses at Acadia University prior to undertaking her laboratory training.[66] Despite a modestly successful year in university, the CSLT was unmoved and held Muriel's application in abeyance, suggesting that she sit for her exam again in the fall of 1944. Steadfast, the CSLT wrote to the Saint John training program that Muriel should "try to complete her matriculation work during night courses."[67] She was to work at the bench and attend class at night to receive her registration, a registration that did little for her employment prospects or opportunities for advancement.

Muriel rejected this. Instead, she continued her education and her working life. She accepted a position at the Moncton Hospital, serving as assistant laboratory technician for fourteen months, before taking a job at the Nova Scotia Sanatorium in Kentville as head technician in early 1945. Her return to Kentville also provided her with an opportunity to resume her university studies at Acadia and she promptly enrolled and completed another half-course in botany. With several years experience behind her, she asked the CSLT "Have I adequate qualifications to receive my registration[?]"[68] This woman, despite feeling rejected by the CSLT merely on the basis of her failure to complete grade twelve, continued her work and her studies. Her own society "felt I was not worthy of my registration,"[69] despite her success passing the exam and completing the training program.

Muriel's experience reveals some of the tensions that accompanied professional formation. Clearly, the failure to be registered did not impede her career but she nevertheless wanted to join the CSLT. Her experience also shows that the new society, which originally embraced laboratory workers regardless of their training, was attempting by the 1940s to restrict its membership through accrediting programs, imposing national examinations and establishing a registry. In part, this was for very concrete reasons such as the negotiated arrangement with the CMA. Less tangible but no less important was the desire to enhance the status and profile of the CSLT.

The impetus to restrict membership, captured in the experiences of Muriel and Helen, contrasts with the CSLT's previous inclusionary strategy. The open-membership policy, as the CSLT's first president Frank J. Elliot declared, was a "necessity if only by virtue of the fact that to have an organization capable of attaining its aims that same organization must be strong in numbers and representation." Such a position was "consistent with British ideals of fair play that those technicians who, through no fault of their own, are not as advanced as others should have representation in an organization such as ours."[70] Similarly the CSLT readily admitted those who specialized in only one aspect of laboratory work, such as serology or histology, upon the presentation of a letter of support from the laboratory director. Elliot did acknowledge, however, that "all future technicians, whether desiring to become general or specializing technicians, should have at least one year of general laboratory training."[71] Initially, the CSLT welcomed workers from small and large hospitals, from Catholic and Protestant institutions, and those who had broad experience in the laboratory or whose work was restricted to one particular area of laboratory tests. There were two membership classes, "active," which included "technicians engaged in medical laboratories," and "honorary," which included "persons otherwise engaged whose assistance and co-operation would be of value in the organising of the society."[72] The promoters of the CSLT had to organize a group with widely different educational backgrounds, work experience, and work conditions. That they attempted to do so at all speaks to the power of the professional image in the health care sector. The early inclusiveness was a practical response for a society struggling to define its membership and its place within the web of hospital workers.

This inclusiveness was also a response to the lukewarm reception the society received from some laboratory directors and physicians across the country, many of whom were less than certain of the need for a national body of laboratory workers. When the CSLT registry was established in 1937, workers received the "registered technologist" (RT) designation. At the inaugural annual meeting of the CSLT, one member asked whether all technicians would be required to register. The president answered, "We cannot force anyone to register." The member realistically replied that "those doctors who do not approve of the Society – their technicians will not be registered."[73] Immediately the society recognized that unsympathetic laboratory directors or hospital administrators would not encourage their workers to join the CSLT and that workers would reject membership for their own reasons. In the late 1930s, entry into laboratory work did not carry with it the expectation of membership in a national society. The professional model for Canadian health care workers may have been ascendant, but workers and physicians alike could avert their gaze.

But many peeked. "Hardly anybody bothered with the Registered Technologists business in those days," recalled one Halifax worker. Apparently, the workers in the Halifax facility reflected the attitude of the director, D.J. MacKenzie. "He didn't care for it one little bit ... because he said that it was a bunch of girls ... He went by what you were like and what you did. He didn't care if you had RT after your name or not."[74] MacKenzie "sure thought RTs weren't any use around the lab, I think that was his theory." In disparaging the CSLT as a "bunch of girls," her supervisor expressed his lack of confidence in the organization in decidedly gendered terms. MacKenzie was also exhibiting a common preference for the apprentice model of training. For him the achievement of the "RT" meant little and certainly did not automatically signify competence. Despite this, the worker decided to hedge her bets and join the CSLT. Some of the workers joined "on our own ... in

Figure 5.2
The First 200 Persons Registered with the CSLT

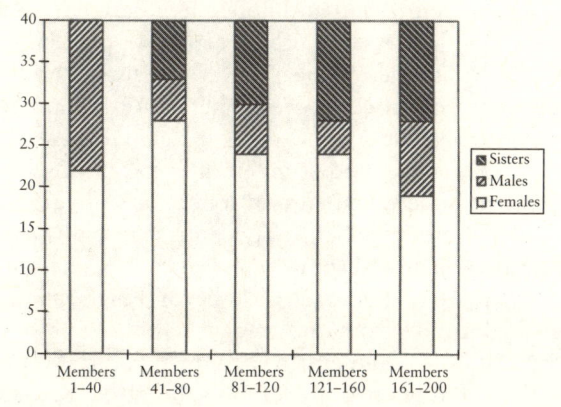

Source: CSLT Registry of Technicians.

case it ever got to be essential."[75] In Saint John, it will be recalled, several workers each year were interested in formal accreditation. Steadily, the membership of the CSLT grew.

The CSLT initially drew on the support of women laboratory workers. Nearly eighty percent of the first two hundred members were women, including forty-one nuns providing service at Catholic hospitals across the country (See figure 5.2). Twenty-one percent of the first two hundred registrants were men. Males were overrepresented in the first cohort of forty members, suggesting that they were quick to join the national body. Conversely, women in religious orders, who represented a fifth of the first two hundred registered, were slower off the mark and underrepresented in the first two groups of forty. It is difficult to know exactly why nuns were slow to join the CSLT, particularly because representatives from Catholic hospitals were prominent in other national health organizations such as the Canadian Hospital Council.[76]

The registry is also a good source of information regarding laboratory workers from the Maritimes and Newfoundland who joined the CSLT in its first decade.[77] From 1936 to 1945, 128 members from the Maritimes and Newfoundland were registered. Registry information was complete for seventy-four of these individuals. They were overwhelmingly single women, although there were seven men. As well, there were twelve nuns. The average age of the members was 27.9, while the average age of the women (excluding nuns) was slightly higher, at 30.1 years of age. Applicants averaged slightly over four years service, but thirty-nine of the seventy-four had two years or less.

A number of factors inflated the age of the applicants. The first was that the CSLT grew initially from persons who were established in their working life and therefore had an interest in a national society. A second factor was that the CSLT imposed a condition of membership of at least twelve months' service. Education also likely inflated the age, by delaying entry into the labour force. Nineteen of the applicants from the Maritimes and Newfoundland had university degrees when they applied to the CSLT, while another nineteen had some university education. Fourteen had undertaken laboratory courses of varying length in a hospital. There were also eight applicants who had nursing education, while another eight had some other education, including business courses. Twenty-three had no education beyond high school, while fifteen combined education from one area with that of another.

Laboratory directors could either encourage application to the CSLT or, through their indifference, undermine the effort to organize. Occasionally, one worker in a laboratory would serve as an agent for recruiting others within a laboratory or a community.[78] The CSLT registry offers indirect testimony to this registration process. The first seven members came from laboratories in Hamilton or Dundas, Ontario. Such clustering of applicants is typical during the early years of the registry. For example, six members from London, Ontario, joined consecutively, while another fourteen joined from Vancouver in a short space of time. Applicants from other cities, such as Halifax, Saint John, Pembroke, Ottawa, and Toronto, were entered consecutively on the membership rolls. The registry suggests, then, that laboratory workers joined the national society in identifiable groups, coming from particular cities or hospitals within those cities. Such a process was undoubtedly governed by the workers' commitment to the idea of a national society, the encouragement of some laboratory directors, and the enrolment of entire classes of students during the course of their training.

With a membership drawn from across the country, it was hardly surprising that the idea of a journal was advanced at the inaugural annual meeting. A professional journal is perhaps the most visible outward sign of a health care organization's commitment to contributing to the advance of health and the production of knowledge. One of the guest speakers at that meeting, Dr Kirk Colbeck, of Welland, Ontario, suggested that the society should have a "bulletin or magazine, typewritten, mimeographed or printed – the form does not matter so long as you have one." Communication was an important ingredient in any national society but it was costly. As Colbeck tellingly reminded the small gathering, "If the doctors can't afford to [maintain communication among colleagues], God help the technicians."[79] There was, however, hope. Colbeck suggested that the various pharmaceutical manufacturers, who supplied the chemicals and reagents for laboratory work, might be willing to advertise in the journal, thereby ensuring its success. The first journal was a mere thirty pages, the second one doubled to sixty-six pages, and then it doubled again to 139. With the third edition, advertisements began to appear in greater number, and in the fourth volume of that initial year, readers were reminded to "whenever possible patronize the advertisers."[80]

By 1940 the circulation of the *Canadian Journal of Medical Technology* had grown to fifteen hundred, from 350 for the inaugural October 1938 issue. The journal was sent to all the members of the society and about eighty pathologists, more than a hundred paid subscribers, eight hundred hospitals in Canada, and twenty-five educational institutions. Subscribers were reported to come from across Canada, although the society conceded that they were "mainly centred in Hamilton."[81] The next year, for example, forty of the sixty-two physicians who subscribed to the journal were from Hamilton.[82]

A tremendous amount of work went into maintaining contact with the membership. In addition to the journal, members regularly received minutes (which totalled 195 pages for the years 1949–54) and the CSLT *News Bulletin*, which was inaugurated in 1951. The minutes and newsletter were mimeographed in-house, on "work nights" when volunteers gathered for five or more hours in the cramped quarters of the head office to collate, staple, fold, stuff, and address the material to the members.[83] The *Bulletin* was bilingual and in 1955, the presidential address included a "few words of greeting" for francophone members.[84] The CSLT had made earlier efforts to reach out to French-speaking constituents. In 1947 it had amended its bylaws to add a francophone director to the executive.[85] Correspondence was often sent out in both languages, and when Ileen Kemp made a "grand tour" of eastern Canada in 1951, a translator accompanied her when visiting Montreal, Quebec, Chicoutimi, and Sherbrooke.[86] Kemp also visited several centres in Nova Scotia during her tour. When she arrived in Halifax, she attended a "lively and valuable" meeting of seventy-five workers and physicians. There were also meetings in Sydney and Antigonish, and Kemp felt the prospects for a Nova Scotian branch of the CSLT were promising.[87]

Despite the initiatives of the national office, attention to French-speaking members, and the perceived potential of Nova Scotia, was the CSLT truly a national society? Data compiled from the registry, presented in Table 5.1, offers insight into the geographic origins of the membership. In the years 1936 to 1945, Ontario, with its close proximity to the national office and numerous hospitals, accounted for forty-one percent of the registered CSLT members; twenty-three percent came from the Prairies, sixteen percent from the Maritime

Table 5.1
Geographic Origins of People Joining the CSLT by Province and Year, 1936–1945

	1936	1937	1938	1939	1940	1941	1942	1943	1944	1945	TOTAL
NF	0	3	0	0	0	0	0	1	0	2	6
NS	0	15	2	8	3	0	7	3	2	12	52
NB	0	6	3	7	5	3	10	4	12	5	55
PEI	0	7	0	3	0	0	0	0	3	2	15
PQ	0	8	1	4	6	15	4	14	4	15	71
ON	13	80	4	25	39	11	25	38	20	53	308
MB	0	7	1	4	2	0	1	13	9	12	49
SK	0	14	6	5	4	4	7	5	7	22	74
AB	0	10	4	3	4	3	3	1	0	20	48
BC	1	33	0	1	1	1	2	3	10	17	69
OTH	0	0	0	0	2	0	1	2	2	2	9
TOTAL	14	183	21	60	66	37	60	84	69	162	756

Source: CSLT Register.

provinces, and nine percent each from British Columbia and Quebec. The remaining persons on the register came from Newfoundland, the United States, or elsewhere. The society was aware that it had work to do beyond Ontario's borders. One correspondent noted that very few of the laboratory workers in Edmonton were registered and that "the general impression has been that the organization is, to quote, 'An Eastern affair.'"[88]

The society did have strong roots in British Columbia, where the national executive appointed George Darling as the provincial representative in 1937.[89] The same year, laboratory workers in Saskatchewan established a provincial branch of their own.[90] In her 1949 presidential address, Ileen Kemp commented on the organizational disparity between east and west. There were vibrant provincial organizations in British Columbia and Saskatchewan and workers in western provinces held their own annual meeting, beginning in 1947. It is clear, however, that when Kemp spoke of the "east" she meant Ontario: "What is Ontario doing?" she asked. The large numbers of laboratory workers in that province made a provincial organization difficult to create. Instead, keen laboratory workers organized "local academies." The academies hosted information sessions, saw workers present papers, and provided a chance for sociability among laboratory workers. Beginning in Toronto in 1949, local academies expanded to other Ontario cities over the next several years and provided a firm foundation on which to build the provincial branch, which was achieved in 1952.[91]

Activities outside Ontario also increased. In 1948 workers in New Brunswick established a provincial branch of the CSLT.[92] By 1950 local activity had reached an unprecedented level. There were three provincial branches in British Columbia, Saskatchewan, and New Brunswick. Calgary boasted an active local group thad had been "in existence for some years," although there was no provincewide organization. Quebec City and Halifax also had active local organizations.[93] Laboratory workers in Manitoba had established a society independent of the CSLT but in 1950 this was reconstituted as a provincial branch of the national organization. CSLT president Joseph Scott marked the occasion by reiterating his desire that "those provinces in which no branch has been formed will soon follow."[94] In 1953 Alberta organized a provincial society, followed by Prince Edward Island in 1954, Quebec in 1958, Nova Scotia in 1960, and finally, Newfoundland in 1961.[95]

EDUCATION

The uneven development of laboratory work throughout Canada, to say nothing of the fragile culture of professionalism, is readily apparent in the struggle to achieve a common education for laboratory workers. The experience of one Catholic sister from eastern Nova Scotia illustrates the circuitous education characteristic of laboratory workers. A sister of St Martha, she attended St Francis Xavier University in Antigonish in 1937–38 and then completed a practical course at the Ottawa General Hospital. From Ottawa, she returned to Antigonish and pursued several summer courses. In 1938 she wrote and passed the CSLT examinations. Periodic trips to Halifax also became a feature of her working life, to maintain her skills in performing tests that were infrequently ordered at St Martha's Hospital in Antigonish and to learn new techniques. Occasionally, she would work "with the staff until I was able to do the test I was interested in as well they could."[96] University education, a period of practical instruction in a hospital laboratory, and informal periodic refresher courses were common features of the educational experience of laboratory workers.

St Martha's was typical of many hospitals in rural communities. From 1938 to 1942, this sister conducted general laboratory work for the 150-bed hospital. During that period, she was the only worker. In 1942 another worker joined her and a third was added in 1948. The laboratory also embarked upon training for its workers, a program approved by the CMA and CSLT in 1947. In 1956 the laboratory underwent a considerable expansion. There were now five departments, expanded from three, staffed by four workers and four students.[97]

The expansion of the labour force, the development of a training school, the presence of students, and the addition of new tests were common features of hospital laboratories throughout Canada in the later 1940s and 1950s, as this woman's experience suggests. Hospitals such as St Martha's expanded, and workers (and students) became responsible for an ever-widening array of laboratory tests. While both the CMA and the CSLT had approved a curriculum, there was no way of ensuring that individual schools were adhering to it.[98] The experience of a student in Antigonish was likely considerably different from that of a student in Toronto or even Halifax. There was also no definite training period. Some courses ran the minimum one year, others took eighteen months or even two years.[99] There were other issues. For example, examiners occasionally refused to accept an approved or standard method that was not used in their particular school – further evidence of the enduring idiosyncratic nature of health care in Canada.[100]

Provincial governments encouraged workers to undertake new training, as the number and range of tests increased. For example, in the early 1940s, New Brunswick sent two senior workers to the Royal Victoria Hospital in Montreal for two weeks' training. One of the workers pursued advanced techniques in hematology, while the other learned new histological methods. The annual report of the Bureau of Laboratories noted that the "laboratory has profited by these girls [sic] added experiences," while the laboratory director, R.A.H. MacKeen, expressed his hope that such opportunities would be continued.[101] Other changes to the day-to-day work of the laboratory were wrought by alterations to the CSLT syllabus of studies. In 1944–45, the Saint John laboratory added lectures in biochemistry to those in hematology and bacteriology. The addition of biochemistry was prompted by the CSLT's desire to have emerging workers familiar with kidney anatomy and function.[102] A registered technician with the CSLT who worked in a doctor's office spent time in Saint John "brushing up and learning the new methods" while her physician-employer was away.[103] Perhaps a University of New Brunswick graduate, also trained in Saint John, best summed up the prevalence of continuing education, suggesting that she "studied a great deal more over the past eight years in the Lab, than I ever studied at College."[104]

While the CSLT apparently avoided the worst excesses of service demanded by nurses' training, this issue did rear its head periodically. One such example occurred in the CSLT's own backyard, at Hamilton's Mountain Sanatorium. This hospital was violating the regulations set out by the CMA. Students were being employed in the laboratory at "a minimum salary" and in place of "qualified technicians." The CSLT concluded that the "students are not qualified to do the work or teach other students under this set up."[105] The executive was also concerned that laboratory workers be fully prepared for their duties. In 1950 they noted a "very disturbing" policy in most hospitals "of requiring the student to accept the full responsibility of night duty after six months of training ... on the whole the practice of requiring students to accept the full responsibility of a qualified technician while still in the training period would seem to be highly undesirable."[106]

Some jurisdictions moved to adopt regulations that would govern student labour. Saskatchewan passed legislation in 1946 that forced hospitals to pay student nurses and student technicians a minimum wage of $18.50 per week, a stipulation vehemently opposed by the Saskatchewan Hospital Association.[107] Other hospitals paid students a monthly stipend, which was graduated so that as students progressed through the program they earned more money. In Ontario, for example, students received a salary during the last five months of training.[108] When the federal government initiated health grants to the provinces in 1950, some of the money found its way to support laboratory training programs and the students enrolled in them. Many provinces offered students bursaries or other forms of assistance, in exchange for a promise of service for a specified period.[109] In other training programs, students did not receive any support and were charged a small tuition fee.[110] There was, then, tremendous variability in laboratory education in the different provinces or even within one province.

Students were not passive victims of poor programs, of course, and complained to the CSLT head office if training schools failed to follow the regulations set out by the CSLT and the CMA.[111] Students also explored educational opportunities close to home, regardless of whether they were approved or not. In the late 1940s the CSLT received inquiries from students who were interested in training at St Catharines General Hospital and St Joseph's Hospital in Hamilton, neither of which was an approved school. In other cases, students who undertook training in programs yet to be certified transferred to approved schools to ensure that they could be registered. Although the CSLT was attempting to restrict membership, it made such concessions because of the shortage of laboratory workers. At the same time, the society recognized that yielding to such requests jeopardized its nascent standards and relations with the CMA. In late 1947 the CSLT executive reaffirmed its resolve that only students attending approved programs would be permitted to register and that transfers would only be permitted "in exceptional circumstances."[112] The CSLT and CMA were the final authority in educational matters and consolidated their authority through the 1950s.

While there was clearly a drive toward standardizing the education of laboratory workers, it is important to remember that in many settings, nurses continued to labour in the laboratory as part of their job. St Joseph's Hospital in Victoria, British Columbia, offered nurses an eighteen-month postgraduate course in medical technology in 1945, although it was the only such course on offer in Canada.[113] Nurses also continued to be instructed in laboratory work in the normal course of their training. At the Owen Sound General and Marine Hospital, student nurses were exposed to the laboratory early in their training. This experience ensured that nurses could adequately interpret the laboratory results that they encountered in clinical care. The nurse would become familiar with the terminology of the laboratory, the classification of disease, and the connection between laboratory results and the etiology and progress of disease. This theoretical work was predictably augmented by a practical turn in the lab. Each nurse at the General and Marine was required to do sixty urinalyses for patients under her care. It was a simple procedure but one that would allow her to correlate laboratory results with the patient's history. The nurse also *observed* the drawing of blood and the tests that were ordered on the samples.[114]

In these latter descriptions, we can discern a shift away from the preparation of nurses for laboratory work. In other words, the training provided to nurses was not intended to prepare them as a laboratory workers, in contrast to the descriptions provided by nurses who worked in the 1920s and 1930s. Rather, the practical and theoretical experience in laboratory technique was to supplement the nurse's education, to give greater confidence, and to provide further insight into disease processes, thereby ensuring some level of understanding of the value of laboratory results for patient care. Despite their continued presence in hospitals across Canada, particularly but not exclusively in rural areas, the laboratory staffed by a nurse was no longer the ideal. Both the CMA and the CSLT articulated a vision of the dedicated laboratory worker in the postwar era and education became a critical symbol of that vision.

Even dedicated laboratory workers, however, did not share a common educational experience. Some workers had a few courses in a laboratory science, while others held arts degrees or graduated from nursing programs. Obviously, the CSLT did much to accommodate the varied educational backgrounds in its early years, leading to a vibrant and healthy organization. Nevertheless, it was greatly concerned with the issue of education, which became one of the society's chief areas of focus in the 1940s. Educational standards, approval of programs, and certification were familiar planks in the effort to professionalize, and, in common with their American counterparts and a broad swath of health care workers, the Canadian society pursued the same goals.

In early 1937 Mountain Sanatorium in Hamilton applied to the American Medical Association for approval of its training program for laboratory workers. The AMA contacted the CMA to see whether it had any objection to the AMA's approving the program. The CMA did not have its own evaluation process, so it raised no concerns. The request, however, prompted the CMA to establish a committee to investigate the education and registration of laboratory workers in Canada. Dr W.J. Deadman, a prime mover in the creation of the CSLT, was appointed chair and Dr G. Harvey Agnew secretary. The rest of the membership, including Halifax pathologist Dr Ralph P. Smith, was drawn from across Canada.[115]

The CMA approval program attempted to ensure that students began laboratory courses with a reasonable knowledge of high-school science and that the programs they entered were accredited in some fashion. The CSLT expressed the opinion that the CMA would not let "every hospital train technicians."[116] When the CMA's committee on laboratory technicians reported in 1939, the CSLT was not disappointed. The CMA committee endorsed the idea of approving schools and set out thirteen requirements for approval, among them the size of the laboratory, amount and nature of the work performed, the qualifications of the instructor, and facilities. The committee decided that schools should be placed "in adequately organized departments of pathology associated with hospitals having at least 400 beds, or in public or other laboratories providing comparable experience." The CMA rejected outright that commercial laboratories had a place in training workers. Course content initially remained the responsibility of the hospital and the laboratory director. The CMA prescribed a twelve-month training program and required twelve months of general training before any specialty training. General training would cover technique in hematology, bacteriology, medical zoology, histology, and pathological chemistry. Specialized training could include advanced technique in serology, bacteriology, or biochemistry.[117] By emphasizing the importance of general training, hospitals could be assured that all graduates possessed a core body of knowledge.

The move toward standardizing what was a complex training system had begun. When the CMA executive met in Winnipeg in June 1941, four laboratories had been approved for training workers.[118] By March 1942, when the *Canadian Journal of Medical Technology* published its first list of approved schools, there were nine.[119] Most of them were concentrated in the eastern half of Canada, but the CMA and the CSLT were confident that the number of schools in the Prairies and British Columbia would increase.[120] Approved schools did grow in number. From nine schools in 1941, the number expanded to thirty by 1946, fifty-eight by 1951, eighty-two by 1956, and one hundred and ten by 1960.

During this same period, laboratory training programs were established throughout the Maritimes. In Nova Scotia the Pathological Institute was approved in 1941, followed by the Halifax Infirmary (1947), St Martha's Hospital (1947), Aberdeen Hospital (1959), the Cape Breton Laboratory of the Nova Scotia Department of Health in City Hospital, Sydney (1959), St Rita Hospital and St Elizabeth Hospital jointly (1960), and Glace Bay General Hospital (1960). There were other options in the region, including the Bureau of Laboratories (1942) and the Lancaster Department of Veteran's Affairs Hospital, both in Saint John, New Brunswick (1950); and the Division of Laboratories, Provincial Health Centre, Charlottetown, Prince Edward Island (1947).

As demand for workers to staff laboratories increased, there were calls to increase the number of training programs. The widespread shortage of laboratory workers during and after the Second World War indicated the need for expanded educational opportunities. One problem was that training programs were enrolling students only to fill vacancies in their own staff, instead of supplying workers for the growing Canadian labour market.[121] In part, the shortage of workers made it difficult for individual laboratories to increase their training capacity. Much of the preparation still depended upon experienced workers passing their knowledge on to students through apprentice situations.[122]

The presence of students undoubtedly aided the work of the laboratory, but the one-to-one ratio of registered workers to students limited the number of learners that approved programs could accept.[123] The prevailing form of apprenticeship training at the bench required that every student be matched with a registered laboratory worker. An education rooted in apprenticeship demanded that a large number of hospitals participate in the training of future workers. The requirement for a one-to-one student-to-RT ratio, however, meant that the total number of students remained small, so that the clinical work would not be disrupted. The small scale of training was also important for the students. Some tests were rarely ordered. The number of students in any one laboratory had to remain small to expose students to uncommon tests and ensure that they performed a sufficient volume of work to achieve both confidence and competence.

Laboratory directors and hospital management in larger Canadian hospitals had to work actively to expand their training programs to meet the labour demand. The CMA's committee on approval noted in 1948 that candidates were plentiful and schools were annually swamped with applicants.[124] If hospitals were to be adequately staffed with laboratory workers, hospitals with the ability to train must do so. Only in this way would smaller hospitals have sufficient staff. Putting it frankly, the CSLT executive commented in 1951 that "this entails an acceptance of the responsibility by the larger hospitals to train technicians for the market as well as for their own needs."[125] That year, there were fifty-four training schools across the country. Despite the addition of six more accredited schools, the shortage endured. A question from the floor of the 1951 CSLT annual general meeting asked whether any more schools were scheduled to be opened. The answer, in spite of the "desperate" need, was not optimistic. The CSLT concluded that most of the major hospitals in Canada had already been approved as training centres. In larger centres, more tests were performed on a daily basis and more workers employed. Thus, the potential for training students was greatest in larger facilities. With only smaller hospitals on the horizon, there was also a diminished capacity to graduate technical workers.[126]

Debates about the entrance standards to training programs accompanied the shortage of laboratory workers in the 1940s. Discussions took on an interesting tenor as some advocated decreased entry standards in order to allow programs to expand,[127] while others debated whether university education was appropriate for laboratory workers.[128] Nor did the move toward standardized education, promoted by the CMA and the CSLT, enjoy unanimous support within the medical and hospital communities. The Registrar of the Ontario College of Physicians and Surgeons remarked in 1948 that a difference of opinion existed among doctors, with "some men preferring students from the Canadian Society of Laboratory Technologists, and others who won't have anything to do with those whom they have not trained themselves."[129] For other physicians, it was a question not of who was doing the training but rather of the content of the curriculum. One student remarked that doctors exclaimed, "I don't believe in general technicians. I want a technician who can do one job well, and I still think only university graduates in science should be employed as technicians."[130] One wonders how prevalent such a view actually was and the evidence on the point is ambiguous. The multiple roles filled by laboratory workers provide some evidence that broad skills were given preference over narrow specialization.

But what constituted specialization? A laboratory in a small hospital required workers with knowledge of a full range of tests and perhaps expected those individuals to assume other duties, for example, in x-ray departments, dietetics, or nursing. Such a worker was highly "specialized," having pursued a variety of training programs and employment opportunities. The multitasking characteristic of laboratory workers before 1950 was anathema to the professional dream of twentieth-century health care workers, each trying to carve out and maintain a sphere of activity in an increasingly crowded and competitive occupational environment. A more elitist view of the laboratory worker likely frowned upon this kind of occupational diversity (one might even say occupational pluralism, although it was pursued *within* the one work environment of the hospital) and this view in all likelihood prevailed in some laboratories. A dedicated pathology laboratory, close to a university, could afford to be more selective in its staffing decisions. If it had sufficient resources, it could also afford to hire persons to conduct a considerably narrower range of tests, in contrast to the smaller hospitals that dotted the Canadian health service landscape.

Standardized education programs, a national exam, and indeed, a national society of laboratory workers dedicated to the advancement of the "profession" cumulatively served to delineate the boundaries of laboratory workers and their job. But despite the vision of the professionalizers, the boundaries remained fluid and continued to be shaped by the labour demands of hospital work. The most dramatic example emerged in Saskatchewan, when Dr W.A. Riddell of Regina inaugurated a combined x-ray and laboratory course for hospital workers in 1946.[131] Riddell cooperated with the CSLT and the Canadian Society of Radiological Technicians to create a corps of workers particularly trained to meet the multitasking needs of smaller hospitals.[132] The first class began on 6 October 1946 and consisted of fifteen returned service people. A second class of twenty was slated for the new year and there were reportedly one hundred and fifty applications for these positions.[133] Students spent three months learning laboratory analyses and another three preparing for duties in x-ray departments. The laboratory training, not unexpectedly, focused on the basic and common tasks, including urinalysis and basic hematology (red and white blood cell counts, hemaglobin estimation, sedimentation rate, and simple staining techniques). Following the six-month training period, students were placed in a hospital laboratory to which either a qualified technician or a laboratory director continued to make supervisory visits for an unspecified period.[134] These workers, given the unwieldy name of "provisional laboratory and radiological technologists," could not become full members of the CSLT until they completed further training.[135]

As the announcement in *Canadian Hospital* acknowledged, the program was not designed to train "fully qualified technicians" but rather so that "the immediate need [for workers] can be met and the technicians can take subsequent instruction and ultimately qualify for certification under the CSLT."[136] Such combined programs became common in provinces with rural hospitals. In addition to Saskatchewan, Alberta also offered a combined course lasting six months, while Nova Scotia and Newfoundland had eight-month programs. The intention of these courses was to prepare workers who could fill positions in smaller rural hospitals and carry out limited duties. None of the programs were offered every year but were a direct response to the labour demands of rural hospitals in an era of expanding hospital services.[137]

In 1950 Riddell offered another educational innovation. He proposed a new training program at Regina College whereby classroom instruction would take place at the college, followed by a practicum in an approved hospital laboratory. This essentially ensured a balance between theoretical and practical education and offered a "more academic approach to training."[138] The time spent on theory and elementary bench training prepared students for a twelve-month hospital-based program, allowing students to enter with a basic knowledge of equipment and techniques and some ability. The CSLT acknowledged that dividing responsibility between the college classroom and the hospital laboratory liberated the latter from the most basic parts of education, which "so burdened" training laboratories. At a time when labour was scarce and large hospitals were already training at capacity, the proposal was sound. It allowed for greater numbers of students to be prepared for practical work, while limiting the time trained staff had to spend teaching rudimentary tasks.[139]

The CSLT recognized there was a need to expand educational opportunities but wished to maintain the balance between practical and theoretical components. This led the national society to reject university-based education. In 1940 the CMA's laboratory committee discussed the possibility of university-based laboratory training but the CSLT shelved the idea.[140] In 1944 the secretary of the CSLT estimated that about fifty percent of those seeking registration held university degrees. Moreover, seven Canadian universities were offering instruction or setting up programs in clinical laboratory technique.[141] In 1949 the national society passed a resolution on educational qualifications using that "every encouragement be made to bring into the Society students who have University degrees and to also encourage present members to pursue their academic qualifications leading towards University degrees but that the present educational requirements not be raised."[142] A balance was achieved. The national society found a way to incorporate university graduates and encourage laboratory workers to further their education in university courses, yet rejected the need to make baccalaureates the standard for entrance to laboratory work.

Nevertheless, university programs continued to develop. In 1943 the CMA general council heard that Queen's, McMaster, and the University of Saskatchewan were planning "more extended courses" for the training of laboratory workers.[143] The University of Saskatchewan offered a certificate in clinical laboratory technique to university graduates who had completed the requisite courses and who spent a year apprenticing in a recognized hospital laboratory.[144] While the program was clearly in keeping with the objectives of the CSLT, it was not an approved course. This prompted Dr G. Harvey Agnew, the secretary of the CMA committee on school approval, to remind the CSLT that only persons graduating from approved schools should be writing registration exams. "This agreement," Agnew declared, "must be clearly understood by all and rigidly followed or the work of [the CMA] Committee loses all significance." Moreover, if the regulations could not be adhered to, the CMA would inaugurate its own registry, thereby subverting the place of the CSLT. The society discussed this letter, noting further that it received numerous applications from programs that were not approved but nevertheless worthy. They agreed to ask program directors to seek approval from the CMA.[145]

Ultimately, the University of Saskatchewan and the CSLT agreed upon a process that saw the exams marked and entered on the university record and then forwarded to the CSLT to be marked for the purpose of registration. When the University of Western Ontario initiated its own course for medical technologists in 1947, the same arrangements were made.[146] Beginning in 1951, Laval University offered a two-year certificate course, with an additional forty weeks of practical training in approved hospitals.[147] Other universities were also contemplating programs of various duration.[148] As each university course was established, the CSLT expressed persistent concern that sufficient practical training be offered, that the internships be with pathologists or other registered training staff, and that the practical work should take place in hospitals with approved training programs of their own.[149]

Debates about the place of university education for laboratory workers also reached the East Coast. In Halifax, Dalhousie University explored a formal course in laboratory technology in the early 1950s and even developed a proposed curriculum. Dalhousie planned to offer practical instruction in laboratories over two summers, followed by an entire year of practical work. The course, which was to be supported through federal health grants, would last for five years and have a special emphasis on courses in biochemistry and bacteriology. Students would earn a BSc and receive the RT designation. During the final year, they would work at the public health laboratory in the Pathological Institute. Students would be paid for this year and for the required summers.[150] The CSLT went so far as to include the Dalhousie program in one of its brochures, suggesting that the university was offering a degree course "made up of three or four years of academic work followed by one or two years of practical clinical training in approved hospital laboratories."[151] However, the program was never implemented. This is not very surprising. In 1951, Dalhousie was still struggling with the presence of a nursing school on campus, established two years earlier on a temporary basis through national health grant funding.[152] The director of the School of Nursing, Electa MacLennan, recalled that a university senate member had difficulty with nurses at Dalhousie and would not countenance more applied programs, such as home economics or engineering. MacLennan recalled that the professor had said "there would be no cookin' and plumbin' on campus."[153] The rejection of laboratory training thus fit a broader pattern of university politics that resisted the presence of "trades" on the university campus, even if they had professional pretensions. University training was also a political issue among laboratory workers. In 1944 Grace Arnold, a CSLT executive member, "questioned the advisability of increasing the years of study necessary to qualify for the work of a medical technologist, in the case that there might arise in the minds of some the tendency to dictate rather than be dictated to, by the medical profession."[154] Her concern was dismissed on the premise that laboratory work exists as a *service* and therefore insubordination was a "rather remote possibility."[155]

Despite this dismissal, the spectre of insubordination arising from university education resonates with the findings of Tracey Adams in her analysis of dental assistants. Dentists expressed their preference for a particular kind of auxiliary worker. Working-class women were suspect and upper-class women lacked the necessary dedication to the work. Middle-class women offered several advantages. They could offer reassurance to a nervous clientele, signal that the dental practice was respectable (and that the dentist would behave accordingly), and discharge other necessary duties, including maintaining the office. Women assistants also worked cheaply and accepted male authority.[156] Dentists were warned against hiring men as dental assistants because of fears that, ultimately, the men would learn basic dentistry skills and enter into competition with male dentists.

Like the dental assistants described by Adams, laboratory workers were to be located as a kind of auxiliary worker within health care, supporting the work of physicians. Such considerations tempered the enthusiasm for university education. There were also underlying tensions with respect to the issue of labour supply. In Manitoba, where the University of Manitoba considered establishing a laboratory course as early as 1945,[157] the provincial branch of the national organization felt that undergraduate education was "unnecessary for all routine laboratory work and that there should be two levels of technologist training."[158] There was a shortage of workers, so most hospital authorities, the CMA, and the CSLT rejected a long period of university education. The opposition to university-based education may have been a practical response to a labour shortage, but many laboratory workers already held university degrees as the Maritime sample suggests. A practical twelve-month course ensured a steady supply of workers, but it also undermined wage claims for these same workers. Defining workers as quickly prepared technical hands without responsibility for determining diagnoses concurrently defined bench workers as inexpensive labour, easily trained and replaced.

RELATIONS BETWEEN THE CSLT AND THE CMA

While both the CSLT and the CMA rejected university education for laboratory workers, the organizations supported the drive for standard education and a national registry. The CSLT and the CMA believed that Canadian hospitals needed a way to recognize "qualified technicians." In this way, potential employers could have "reasonable assurance" that those registered would have a basic knowledge of laboratory procedures.[159] "The technician ... plays an important role in the chain of diagnosis and treatment of the sick as well as in the field of preventive medicine," added the CMA committee on laboratory technologists. "The importance of these workers has steadily increased during the last forty years and his [sic] status has at the same time improved. Moreover women have taken an increasing part in this branch of medical work. With expanding usefulness the necessity of higher preliminary education and more elaborate training has developed."[160]

To ensure that laboratory workers were adequately prepared to assume their duties at the bench, national standardized examinations were implemented for admission to the registry. The CSLT took a decentralized approach to examinations. A local laboratory director interviewed interested applicants and administered the examination. Exams were scheduled twice yearly and consisted of a practical exam, which was not to exceed one hour, and a written examination of between three and four hours. Both portions of the exam were weighted equally and the pass mark was seventy percent. There was also a provision for workers specializing in particular segments of laboratory work. If the individual had one year's experience, a letter supporting that person's application from the laboratory director was sufficient.[161]

The CSLT and the CMA clearly wanted to impose national standards on laboratory training, registration, and work. The CMA in particular carefully guided the effort to establish the registry, set exams, and approve education programs. The creation of the registry, under the

CSLT's responsibility, is an important symbolic achievement. It as[?] the employer that those registered, even if they have acquired their e[du]cation in different settings, share a core knowledge and set of comp[e]tencies. In this way, the idiosyncratic education of laboratory workers, replete with programs in hospitals, colleges, and universities, could be brought to order.

The CMA, through its laboratory committee, also kept an ever-watchful eye on professionalizing laboratory workers. The CMA's laboratory committee carefully considered every school that applied for approval to train laboratory workers. Schools were initially sent a questionnaire, which was augmented by the personal knowledge of committee members regarding the ability of the school to conduct a rigorous education program. Occasionally, schools were discouraged from seeking approval on the basis of the personal knowledge of the CMA committee members. On other occasions, the committee requested more information from schools that applied. If applications were received but committee members continued to harbour doubts, an inspection was carried out. Applications from inferior programs were rejected:[162] expansion of the labour force would not occur at a cost. Such a high-minded stance was probably easier to pursue in an occupation that still did not require certification from its workforce. After all, rejecting schools did little to diminish the labour supply. Unapproved laboratories could still offer training to fulfill their own needs and these workers could still find positions in laboratories across Canada.

With the growing number of approved education programs, the CMA continued to monitor the situation closely. In 1944 it expressed concern over the lack of training standards. The CSLT admitted the first 275 members without any examination whatsoever, in an effort to "give the Society a start." The CMA committee erroneously believed that by the mid–1940s the laboratory society was only allowing persons enrolled in approved schools to sit for examinations. In fact, the CSLT was still permitting *any* worker to sit for examinations, regardless of when they completed their training or whether they met a minimum standard of education, as the earlier example of Muriel vividly illustrated. The CMA believed it had a verbal agreement with the CSLT that after a specified date, only graduates from approved training programs would be permitted to write examinations and only those who passed the exam would be entered into the register. There was some disagreement between the CSLT and the CMA regarding the date, however, as there was no written agreement. The CMA committee decided unilaterally to establish the date again at 1 July 1945, after which time only those who passed the examination would be entered into the registry.[163]

The struggle over exams and registration was only the most overt example of medicine's interest in controlling the labour supply for the ever-increasing number of hospital laboratories across Canada. Indeed, through the laboratory and education committees, the CMA was establishing control over the education and training of laboratory workers, while the CSLT was still struggling to attract members and define a role for itself among health care workers. The CMA delegated control of the national registry to the CSLT, but it was clear that the CMA would ensure that the standards that *it* defined were enforced by the laboratory organization. The CSLT may have been striving to create a national voice for laboratory workers, but at best it was a voice singing in concert with professional medicine, and at worst, it was only part of the chorus. Replicating the hospital hierarchy, the CSLT would be subordinate to the CMA.

CONCLUSION

It is easy to imbue the creation of national professional organizations such as the CSLT with symbolic significance. After all, such societies provide a ready date from which to begin a narrative of professional identity and uplift. It is also apparent that the CSLT acquired many of the hallmarks of a "professional" organization, including approved training schools, national examinations, a registry, and a professional journal. The CSLT experienced dramatic growth in the late 1930s through the 1940s, growing from approximately two hundred members in 1937 to close to fourteen hundred by 1950. It is also clear that the CSLT shaped, in part, the development of laboratory work. Together with the CMA committee on laboratories, the CSLT approved schools, created and administered exams, and controlled the registry (itself an important symbolic achievement). In some ways, the CSLT developed all the organizational hallmarks of a profession. But such evidence offers only limited insight into the broader place of laboratory work within health care and reveals the limits of the search for professional traits, discussed in the introduction. It also reaffirms the need to return to the archives to examine the working lives of health care workers and the material conditions under which they laboured.

A closer examination of the working lives of the membership reveals a more complicated story. There were many portals through which one could enter laboratory work and these endured. The multiple routes to the laboratory were not the vestigial limbs of an earlier era but rather a realistic response to the needs of small or large hospitals, or rural or urban facilities; in short, the labour needs of the different components within the emerging health care complex. Laboratories were increasingly important, to public health, clinical care, and teaching and rese Smaller or specialized hospitals recognized the growing importance laboratory analyses in the clinical setting, but they had limited resourc and therefore sought ways to fill the necessary positions by combining laboratory work with other services.

The demand for laboratory labour demanded that training programs be relatively brief, with an emphasis on practical components. On the one hand, this led to innovations like those of W.A. Riddell in Saskatchewan. On the other hand, it also led both the CSLT and the CMA to reject university education. Laboratory workers for the most part continued to train in hospitals, which reinforced their position vis-à-vis physicians. Physicians continued to play a role in educating laboratory workers (at least they had formal responsibility as the laboratory directors) and it was physicians who examined them. As we shall see in the next chapter, physicians also attested to the moral character of applicants to the CSLT. The education of laboratory workers facilitated their *inclusion* in the health care system in very particular ways, as workers who primarily supported the work of physicians and who could, when required by hospital administrators, work in multiple departments. Moreover, the rejection of university education was linked in part to the idea that laboratory workers were not to challenge the authority of physicians.[164] The debates about education at best reinforced the idea that laboratory work did not require lengthy preparation and, at worst, marked their training as second class, even if the programs were approved by the CMA.

It is hard to define laboratory work precisely. So much depended upon the particular setting, where often the boundaries between occupational groups became blurred. Training shifted to respond to local needs, suggesting the ongoing importance of place in any analytical framework, a theme that will be pursued in the next chapter. Rather than stable identities and specialization, one can see fluid roles and the multitasking that prevailed among the membership. This presented challenges to the nascent CSLT, particularly in establishing its membership rules, registration criteria, and debates about education. The ideological drive to professionalize ran headlong into the material reality of hospitals, which sought ways to economize by adding duties to existing staff or assigning multiple tasks to new staff. The professional identity of laboratory workers was not stable but rather fragile and fluid. This uncertainty, as we shall see, is reflected in the idealized portrayal of the laboratory worker.

6

Recruitment, Mobility, and Wages

In 1931 the Canadian Tuberculosis Association (CTA) and the provincial government of Prince Edward Island were attempting to recruit personnel for the newly established public health department.[1] But, as in New Brunswick and Nova Scotia, there were limits to what the province could afford. The development of public health in PEI depended upon funds from nongovernmental sources. In 1930 the PEI government spent a mere twenty-four hundred dollars on public health and the total expenditure on the island, from all sources, was just under nineteen thousand dollars.[2] The CTA brokered a deal with the Canadian Life Insurance Officers Association to provide an initial grant of fifteen thousand dollars per year for five years, beginning 1 July 1931. These funds provided a salary for Dr Benjamin C. Keeping, who became the assistant medical health officer, and supported his postgraduate training. The CTA's money also supported sanitary inspectors in Summerside and Charlottetown and paid for office staff for the chief health officer. Red Cross nurses, who had provided a great deal of service to PEI, were transferred to the staff of the new provincial board of health and additional nurses hired. Furthermore, a new public health laboratory was established in the sanatorium, which was also new, and staffed.[3]

In early 1931 a number of candidates were being considered for a position in the laboratory, including a female doctor who was not in practice because of poor health (she had tuberculosis), a graduate from Dalhousie, described as young and "less experienced," and Miss Marion Merry, who resided in Toronto. Merry had trained in Reading, England, and worked for five years at the Toronto General Hospital, where she was responsible for bacteriological and pathological tests. The successful candidate would be trained by D.J. MacKenzie at Halifax's Pathological Institute. Dr P.A. Creelman, the health officer for PEI, acknowledged to MacKenzie in March 1931 with remarkable frankness that the person selected would have to work very independently because "I do not think there will be anyone here qualified to give her much supervision."[4] In February 1931 CTA executive secretary R.E. Wodehouse had expressed the same view, writing to the prospective candidate, "You will mostly be your own boss." He added, "You will be able to do some work for physicians, such as blood work and so on, for which it might be possible to arrange fees [on a] private work basis." To further entice the candidate, Wodehouse noted that meals were to be provided, that the public health nurses would be available to assist in the laboratory when required, and that the work and location were both "desirable."[5] Correspondence was exchanged through February but Marion Merry ultimately declined the position. Writing to Wodehouse on 22 February, she cited "purely private reasons of a fairly serious nature" and the distance of the move from Toronto, adding, importantly, that she could not afford "to remain in idleness until the middle of the summer when the appointment commences."[6]

The search for a staff person continued. Wodehouse contacted Dr Robert Defries of Connaught Laboratories in Toronto asking whether he knew of a suitable candidate.[7] After further consultation, this time with Dr John FitzGerald, another of Toronto's leading public health figures, Wodehouse decided on another path. Rather than try to recruit a trained technician from Toronto or another centre, a nurse from PEI would be sent to Halifax to train at the Pathological Institute. Wodehouse wrote to Keeping that:

I discussed the matter in Ottawa Wednesday with Dr Fitzgerald at considerable length and he thought we would be well advised to take this course, because, if the nurse has friends in Charlottetown she would be socially at home there. He felt that she likely would be more suitable than someone strange to the Island whom we might find to take there, and he also felt that in the long run ... a contented person with long service would be of greater advantage to us than an over-efficient person to begin with who might become restless and leave us at the end of nine or twelve months.[8]

Staffing decisions were shaped by a wide range of variables, including home or place as in the above example, religion, or local employment conditions.[9] Skill and efficiency were not the only considerations when filling positions in the emerging health care complex.

Esther Stevenson was a graduate of Prince of Wales College, who taught school on Prince Edward Island for four years. In 1922 she began her nursing education in Massachusetts. She undertook further training in mental health and tuberculosis work in New York and Northampton, Massachusetts. In early 1931 she wrote to Creelman that she had "recently had a course in Tuberculosis Nursing at Trudeau Sanatorium [New York] ... which included Laboratory work, urinalysis, staining and recognizing tubercle bacilli, with lectures in bacteriology."[10] While her preparation was modest, she was an Island nurse willing to undertake further laboratory training.

Stevenson would also work for less money than a more fully trained worker. Indeed, Creelman proposed paying her one thousand dollars a year, whereas Merry commanded fifteen hundred.[11] The CTA was willing to pay eighty dollars a month to cover Stevenson's maintenance while she undertook eight months' training at the Pathological Institute in Halifax. This would provide her with a working knowledge of Kahn tests for syphilis (which had replaced the Wassermann), a range of bacteriological tests, and familiarity with the latest in milk and water testing. Blood chemistry was not considered a priority. "I would like her to have it," wrote Creelman, "but if she can do the ordinary examinations required in public health work we would be satisfied."[12]

The selection of Esther Stevenson is instructive for a number of reasons. First, there was the important role of the Canadian Tuberculosis Association in providing funding. Such ventures were a familiar component in the Maritimes during the formative years of public health. The Rockefeller Foundation, for example, provided over $675,000 from 1920 to 1940 to support a variety of health initiatives in Nova Scotia, though most of the money went to Dalhousie University's medical school.[13] The pattern of broad cooperation that prevailed in health care, including federal-provincial initiatives, philanthropic and volunteer activities, Maritime-wide strategies, provincial-municipal cooperation, or coordinated services across municipal boundaries, is an intriguing feature of health services in the first half of the twentieth century. New health care initiatives in the Maritimes depended upon such broad support; where it was absent, innovations withered. For example, the Rockefeller Foundation sponsored a rural health unit in New Brunswick in the early 1920s, spending approximately $45,000. When the external funding ran out, the municipalities were unwilling to shoulder the burden and the effort collapsed.[14]

There was also the important question of attempting to hire a person with local connections, both to facilitate recruitment and to enhance the possibility of keeping the person for a substantial period of time. Stevenson's appointment reveals that staffing decisions were not made solely on the basis of qualifications. She was certainly not the most qualified person and PEI's public health officials, including Creelman and Keeping, recognized this. Equally enlightening is the variability of the women considered for the position in the public health laboratory. There were two candidates with laboratory training, a number of nurses, and two female physicians. All were thought to have the required skills to take charge of the lab, even in the absence of a supervisor. After all, the successful candidate, according to the physicians, was to be "her own boss." Moreover, relief and assistance in the laboratory were to be provided by the public health nurses.[15] In the end, Esther Stevenson never assumed the position in the laboratory, despite undertaking training in Halifax. Instead, she opted for marriage while in the midst of her training.[16]

SALARIES AND MOBILITY

The efforts of the Prince Edward Island government and the Canadian Tuberculosis Association reveal a number of important issues related to salaries, retention, and occupational mobility for laboratory workers. Like many workers, those who laboured at the bench endured wage reductions during the depression. However, there was considerable variability among workers in different employment situations and settings. Salaries were restored for workers in Nova Scotia's public health section in 1935, but the gesture did not extend to workers in the pathology section.[17] There were other differences in salary as well. Public health laboratory workers in Halifax were generally paid more than those who worked in the pathology section, and workers did not achieve parity until 1 September 1939.[18] In New Brunswick, the passage of the 1944 Civil Service Act, which standardized starting wages and scheduled increments, was hailed by the Bureau of Laboratories as a measure that "should enable us to attract and keep technicians." "More than anything else," the Bureau of Laboratories' annual report mentioned, "it has produced a feeling of stability and permanence which was lacking before."[19] Attrition through marriage, the lure of the United States, and better-paid opportunities elsewhere in Canada complicated the issue of staff retention for Maritime laboratory directors.

The CSLT, still struggling to carve out an identity in the 1940s, also became concerned with the ebb and flow of its membership. Among the most important issues for the CSLT, and one in which the Maritimes had a particular interest, was salary standardization.[20] Members from New Brunswick argued that the health department was not only falling behind other provinces but also behind the rates paid at "smaller hospital laboratories" within that province. Evidence from Nova Scotia supports the claim.[21] A combined technician earned eighty dollars a month at Dawson Memorial Hospital in Bridgewater, Nova Scotia, in 1941. The next year, the technician relocated to Halifax to pursue general laboratory work at the Pathological Institute. Her salary remained eighty dollars but she no longer received her meals, room, or laundry as she had in Bridgewater. Thus, at a glance, her job change brought a salary reduction. This was offset, however, by her ability to return to her parental home.[22] Others were not so lucky. In 1941 a technician with three years' experience earned seventy-three dollars a month in the Saint John laboratory, with no meals or room.[23] In 1942 the New Brunswick laboratory director believed the availability of higher salaries elsewhere meant that New Brunswick could only retain "local girls."[24]

In 1942 the Canadian Hospital Council studied survey responses from 230 "representative" hospitals across the country and then divided them into categories based on bed capacity.[25] The sample was extremely limited, for there were well over seven hundred hospitals in 1942[26] and much of the information was based on very few respondents. In some instances only one questionnaire was returned. The data, presented in Table 6.1, were adjusted for maintenance such as meals and rooms and also excluded the very small salaries earned by religious orders. Hospital workers' wages showed tremendous variation, with hospital size and setting among the critical determinants. For example, kitchen staff in the smallest hospitals were paid as little as ten dollars a month in one Saskatchewan hospital, while a hospital in Quebec paid forty dollars for the same work. Only two hospitals in the twenty-six- to fifty-bed range reported having housekeepers, and their salaries were eighteen and sixty-five dollars respectively. Local economies, the variety of demands on workers, and the patient load of the hospital established what a worker would be paid. There is a need, then, to situate workers in their regional context. Hospitals in this period were constituted not only through advances in medical science, technology, or therapeutic innovations. They were also structured by local issues, including municipal taxation rates and citizen participation, among others.[27] While hospital development occurred across Canada,[28] individual hospitals showed tremendous variation depending on the institution (and its clientele), its setting (rural or urban), and its size. Salaries for technicians in hospitals of between fifty-one and one hundred beds ranged from sixty dollars with full maintenance to one hundred and twenty-five dollars with one meal. Regional averages, however, showed only slight variation. The Maritimes averaged $75.45, Ontario $76.25, and the Prairies $73.50.

In 1943 Dr Arnold Branch, the director of the Bureau of Laboratories, submitted a letter supporting increased wages for New Brunswick laboratory workers. The letter, which was read into the CSLT minutes, recorded the inadequacy of wages in Saint John and the resulting difficulties of securing and retaining staff: "From a purely selfish point of view it is becoming increasingly difficult for me to retain our girls or employ newly trained staff. We have never been able to engage other than local girls who can live at home, as the initial salary we pay is not enough for anyone dependent on boarding out and the maximum reached after six years is about what the minimum should be."[29]

Table 6.1
Monthly Wages for Selected Canadian Hospital Workers, 1942

Job Title	*Number of Beds*				
	Up to 25	*26–50*	*51–100*	*101–200*	*201+*
Superintendent	$73.50	$103.50	$138.75	$174.00	n/a
Assistant Superintendent	$70.00	$76.50	n/a	$103.75	$287.66
Business Manager		$118.00	$139.00	n/a	$241.00
Accountant		$74.50	$77.50	$96.50	$166.00
Nursing Superintendent	n/a	n/a	$125.00	$126.56	$176.00
Assistant Nursing Superintendent				n/a	$118.50
Instructress		$70.00	$84.45	$95.00	$113.55
Night Supervisor		$67.00	$73.00	$82.00	$96.50
OR Supervisor		$72.30	$76.00	$86.90	$104.40
Obstetrical Supervisor		$71.66	$71.50	$78.90	$96.50
Other Supervisors			$62.66	$71.00	$79.50
Graduate Nurses	$53.50	$61.85	$57.00	$55.90	$59.35
Technicians		$78.75	$75.45		
Radiology Assistant					$213.45
Radiology Technician				$74.60	$82.30
Radiology Chief Technician					$143.25
Pathology Technician				$67.00	$83.25
Medical Record Librarian			$70.00	$68.20	$92.90
Dietitian		$64.80	$66.90	$80.95	$104.45
Kitchen Help	$23.90	$23.30	$23.20	$25.85	$43.70
Housekeeper	$30.00	$41.50		$40.45	$70.30
Janitor	$43.20	$72.25	$82.50		
Maids	$16.95	$20.70	$23.25	$19.35	$26–32
Orderlies	$35.00	n/a	$46.00	$48–61	$56–66

Source: *Canadian Hospital* 19 (October 1942): 44–50; and (November 1942): 34–40. Blanks in the table indicate that the job described did not exist at the reporting hospitals. Where insufficient data were reported (even though the position was present), n/a has been used.

Branch renewed his complaints about New Brunswick salaries the next year in Toronto.[30] In 1944 his Nova Scotian counterpart, Dr Ralph P. Smith, joined him. Reporting to the general council of the CMA, Smith requested a survey of technician salaries across the country. The results suggested that workers in the Maritimes were paid at a "much lower" rate than their counterparts elsewhere.[31] CSLT president Frank J. Elliott wrote to J.A. Doucet, the New Brunswick minister of Labour and Health, that he found it "difficult to understand how technicians in

New Brunswick can be expected to clothe themselves and pay living expenses, etc., on the salaries which they are now receiving. I do not know how the Directors of your laboratories can keep their trained staff, or replace them when they leave, for I am sure you must be losing a number of well trained technicians."[32] A Saint John worker wrote in the early 1950s that new recruits were discouraged from laboratory work because of the inadequate wages and suggested to her confidante that "they weren't like us – we had to pay board in the city and buy everything with not a penny paid."[33]

It is clear that the low salaries and limited opportunity for advancement were hindering recruitment and retention of laboratory workers in the Maritimes during and after the Second World War. The 1942 *Canadian Hospital* survey (Table 6.1) reveals remarkable variation among laboratory workers, although regional differences were seemingly not significant. Given the exasperation of Maritime laboratory directors and the small survey sample, it is likely that the survey simply underreported regional discrepancies. As noted earlier, there was the lure of the United States and other parts of Canada. Even within a province, some hospitals paid better than others. Opportunities were plentiful and laboratory workers frequently pursued better positions. Neither the CMA nor the CSLT thought they could do anything about the wage disparity across the country. Nevertheless, the head office of the CSLT suggested that laboratory workers enduring low wages "leave and take [other] positions at higher salaries thus making it imperative that ... [employee] salaries be raised in order to bring in an adequately trained staff to do this work."[34]

While local economies, patient load, and the variety of laboratory analyses performed all affected worker salaries, other factors helped Maritime hospitals retain staff. There was a desire to stay close to home for some, which allowed individuals to contribute to the maintenance of the family or, at the very least, to avoid paying room and board. Many laboratories, as directors were quick to point out, employed mostly "local girls." While the directors viewed this as a sign of their inability to recruit outsiders to the positions, for some of the women the option to work close to home was a happy one. As Kathleen Harvey, superintendent of the Middleton Hospital in Nova Scotia, reminded advocates of university education for all nurses in 1937, small hospital-based programs offered young women the chance to become nurses "almost within the shelter of their homes."[35] Just as recruitment strategies considered variables such as where one was born or the bonds of kith and kin, decisions by workers to stay or leave positions were not exclusively motivated by better wages.

By the 1930s, when significant additions were being made to the staff in both Halifax and Saint John, laboratory work was overwhelmingly performed by women. Yet how exactly were women recruited to work at the laboratory bench? For some workers, the laboratory was one of a range of options. Such things as interest in the work and ability, neighbourhood information about available positions, wages, and marriages shaped the way women came to work at the bench. The social relations of gender, age, ethnicity, and religion also created essential connections that could lead to employment. Laboratory work, after all, did not enjoy the same visibility as other kinds of "women's work," such as nursing, teaching, or retail sales.[36] Yet laboratory work was respectable employment and it became a significant niche for women who wanted to work in health care (or, more generally, public service) but who did not want to become nurses.[37] For some women who lacked the means for an extended period of education, the shorter training period was attractive. For others who were exposed to laboratory technique in university, laboratory work presented an opportunity for them to pursue their interest.

Of course, not all women enjoyed the same opportunities. Rose Phillips was certain that she would be relegated to teaching, a career prospect she did not find particularly enticing. She returned to her native Bermuda following graduation in May 1931 but went back to Halifax and her job in the laboratory in late August.[38] Phillips recounted that she did not actually apply for her position but rather responded to an informal offer of employment. It was the break she was looking for, not having much desire to return to the limited employment opportunities in Bermuda.[39] Another woman, who began working in 1940, recalled that her parents thought that laboratory work presented a good opportunity for their daughter. Nor did the rest of her family, friends, or neighbours find her choice of job to be unusual. She recalled "it was a service for people," suggesting that it did not deviate too much from the prescribed roles for working women.[40]

Ellen Robinson, who joined the laboratory staff in 1936, articulated the ways in which laboratory work differed from nursing. Robinson rather enjoyed nurses' training and even liked working in hospitals. But opportunities were few and she found employment as a private duty nurse shortly after graduation. She recalled how "sitting up with people with strokes, watching them die" held little appeal for her.[41] In the laboratory women found satisfying and remunerative work in health care without having to deal with sick patients on a day-to-day basis. Increasingly Robinson looked for new opportunities, eventually securing a laboratory position. Six decades later she recalled that her parents "didn't think anything" of her decision to change jobs.[42]

Robinson began to make forays into the laboratory between assignments as a private duty nurse. Given her interest and aptitude for science, as demonstrated during her final year at Dalhousie, she thought that laboratory work might provide a satisfying alternative career to nursing. Robinson completed a short course of training at the laboratory and found that she particularly enjoyed working with pathology specimens. When Rose Phillips left to be married in 1936 the laboratory director, Ralph Smith, recruited Robinson.[43]

Robinson's entry into laboratory work was complex and multifaceted. She had an extensive, though orthodox, education, one that took her to the Halifax Ladies' College, through a domestic science program, then to Dalhousie University, and finally, to nursing education. A botany professor recognized her aptitude for science and offered some encouragement.[44] Robinson's education, however, was only one component of her entry into the field of laboratory work. She had been introduced to it before her nurse's training by Dr A.G. Nicholls, longtime director of the laboratory and her neighbour on South Street in Halifax. Personal connections also played a key element in her actually securing the job. In the summer of 1936, just prior to joining the laboratory staff, she worked for a summer at a camp operated by the warden of Shirreff Hall, a woman's residence at Dalhousie. The warden was an aunt of Dr R.A.H. MacKeen's, assistant professor of pathology at Dalhousie. Finally, it was the impending marriage of Rose Phillips (and the resignation that would follow) that finally secured the position for Robinson. Education, personal knowledge, and familiarity, together with the constraints on working women generally, all combined to lead Robinson to laboratory work.

Many women combined university courses with work in university laboratories. One Halifax woman had five and a half years experience by the time she applied to the CSLT in 1944 and it is clear that she worked in the laboratory throughout her university education. A Fredericton woman "studied two afternoons a week" at that city's Victoria Public Hospital, where she did routine laboratory work. Another applicant reported that she worked in the biology laboratory at Acadia University in Wolfville, Nova Scotia, for two years during her studies.[45] Exposure to laboratory technique while at university and the "freedom" to volunteer at local laboratories was an important entrée to the work for many women.

There were other paths to the laboratory bench. Margaret Robins, for example, was appointed following the resignation of Deborah Henderson in early 1931, while concurrently Virginia White was promoted to the position of senior technician.[46] Some women, like Margaret Low, were well known to the medical community and acquired a high degree of competence in their work. But there were other means of securing a position. Louise Gowanloch worked at the public health laboratory for six months in the mid-1920s. She was the wife of James Gowanloch, a Dalhousie zoology professor hired in 1923. Louise Gowanloch was hired, like many of the staff additions, to perform VD tests.[47] In 1924 she corresponded with hospital superintendent W.W. Kenney about the possibility of finding work at the Victoria General Hospital.[48] It was unusual but not entirely unheard of for a married woman to work in the laboratory before the Second World War but the Gowanlochs were not a traditional Halifax couple. They lived apart for much of their time at Dalhousie. Louise studied medicine in New York, but even when she lived in Halifax she lived in a boarding house. In 1929–30, after finishing her medical degree,[49] Gowanloch took a one-year contract with Dalhousie's biology department. Her marriage ended in September 1930, in a divorce scandal that rocked the university.[50] There were other exceptions, such as Mrs M.A.H. Swim and Mrs Dobson, about whom we know very little,[51] but laboratory workers were overwhelmingly single women.

Edna Williams "always knew" that she was going to work. Williams remembers being "desperate for a job." Searching for work while a university student, she had her "name in everywhere. Banks and the archives and everything I could think of … Simpson's, Eaton's." Williams, who started at Dalhousie in 1934, applied for various jobs every summer, including for a position at the laboratory. She was eventually interviewed by D.J. MacKenzie and was successful, in part because she had a university degree. This also translated into a monetary reward for Williams. Workers without degrees received sixty-five dollars a month, which was a typical rate for clerks in government. But people with degrees in the public health laboratory earned eighty dollars a month.[52]

While some women pursued laboratory work as an option to other kinds of employment, it was also respectable work for daughters of the upper class. Edna Williams recalled that "society girls would work in the hospital labs, the debutantes."[53] Other women found the opportunity to change their career paths liberating. Laura Piers was educated in the Saint John facility for fifteen months, from August 1944 to November 1945. Shortly after completing her training, she assumed a position at Moncton City Hospital, serving in a supervisory position for one month. Subsequently she worked at the Blanchard Fraser Memorial Hospital, a Department of Veteran Affairs hospital in Sussex, New Brunswick. In 1947 she took a hematology course at the Thorndike Memorial Laboratory at the Boston City Hospital and from there she went to the Toronto General Hospital, working in hematology and biochemistry from 1948 to 1959. Piers later worked at St Thomas Hospital, London, England, at the Royal Victoria Hospital in Montreal, and then back to St Thomas. She finally returned to Halifax as the senior technologist in the hematology section in November 1964.[54]

While her travels were more extensive than most, many workers' careers exhibited a similar pattern of geographic mobility as they pursued opportunities across Canada. Robert Mitchell arrived in Canada in 1930 and went to work at the Queen Alexandra Sanatorium in London, Ontario, where he stayed until 1937. From 1937 to 1942, Mitchell worked for the International Nickle Company Hospital in Copper Cliff, Ontario, where, in addition to his laboratory work, he served as an assistant x-ray technician. Industrial work was not uncommon. Ann Macauley was the senior laboratory technician at a munitions company hospital in Cherrier, Quebec, during 1942 and 1943. There she performed routine tests in hematology and biochemistry and conducted research projects on employees who were exposed to trinitrotoluene, or TNT. In 1943 she joined the staff of the Moncton Hospital and, after a year off, went to the Moncton Tuberculosis Hospital, where she served from December 1946 to the spring of 1950. She later relocated to Ontario, where she worked for the Ottawa Civic Hospital, a physician in private practice, and the Bell Telephone Company.[55]

Christie Hart had a similarly diverse career. She had completed four months of laboratory training and six months of x-ray training at the Montreal General Hospital. She then went to the Miramichi Hospital in Newcastle, New Brunswick, where she worked as an x-ray technician. Hart returned to Montreal, where she worked in Dr R.F. Kelso's Macdonald College office. She stayed in that position for eight years doing x-ray, laboratory work, basal metabolism, and shortwave diathermy, wherein electromagnetic currents are used to generate heat in body tissues. Finally, in 1943, she joined the staff at the Royal Canadian Navy Hospital laboratory in Halifax.[56]

Laboratory workers, in common with many daughters and sons of Maritimers, left the region as well. Maureen Noonan laboured at the public health laboratory in Charlottetown from 1943–45 testing sputums for the sanatorium and performing an array of public health tests. In 1945 she went to Kingston General Hospital and worked in the laboratory as a junior technician from 1945 to 1947. She then entered the Canadian Red Cross Blood Transfusion Service, serving in the Edmonton and Calgary depots (1947–48), Vancouver (1948–49), and Hamilton (1949–52). In the last position, she served as senior technician. Noonan then went to New Westminster, British Columbia, working in the Royal Columbian Hospital. In 1955 she returned to the Red Cross, working at the national headquarters as a technical consultant.[57] In slightly more than a decade, her career had taken her across Canada to seven different cities. Such geographic mobility was common among laboratory workers in the years following the Second World War, as hospital diagnostic services expanded.

For many laboratory workers, their university experience exposed them to scientific apparatus and nurtured their interest in laboratory technique and undoubtedly their skills. University also provided many, particularly the Halifax women, with personal contacts that led them into laboratory careers following graduation. Labour in the laboratory provided women with an opportunity, albeit limited, to pursue a career in science. Despite the very real limitations of such a career, laboratory work should not be viewed as a kind of occupational ghetto for women. Some found that work in the laboratory offered a chance to travel, whether to learn new tests or to pursue new employment opportunities. To reiterate, it was stable, remunerative, and respectable work that provided educated women with an outlet to pursue their interest in science or to escape from the demands of caring for the sick or the dying (as in nursing) or children (as in teaching). In a period of limited opportunities for women, laboratory work was a welcome alternative.

DEMAND FOR WORKERS

If some of the women who entered laboratory work found it satisfying, there were also a fair number of opportunities in the field. As we have seen, positions were available in many parts of Canada and beyond. With the Second World War, jobs for women expanded generally and the demand for technicians increased both in the armed services and at home.[58] In Halifax one woman recalled that vacant positions were filled with "older [workers] that had left and they also, of course, kept the married ones [whose] husbands were overseas."[59] Beginning in early 1940, the CSLT began exploring ways in which its membership could best serve the war effort. The society created a military service register to identify interested members. If they enlisted, laboratory workers were admitted to the armed services as privates, in sharp contrast to the nurses who joined the medical corps as officers.[60] The lower rank was justified in the military's view because the duties in military hospitals consisted of only routine tests in hematology, biochemistry, bacteriology, and the ubiquitous urinalysis.[61]

For its part the CSLT was committed to defending the standards it had established in the late 1930s. Recognizing its fragile and emerging professional recognition, the CSLT resisted a military classification as a trade. Laboratory workers in the military were carrying out the same duties they performed in civil life, the argument went, so why would a civilian professional group be downgraded to a trade in the military? As a correspondent to the *Canadian Journal of Medical Technology* ably pointed out, "In our enthusiasm to help our country we must not overlook the effects of our actions on the profession after the war is over." The CSLT must do "everything in their power to prevent the lowering of the standards of our profession."[62] Professional standards were, however, still being negotiated.[63]

By 1942 the national office estimated that two-thirds of its male members were serving in the armed forces. Male members were "conscientiously accepting their responsibility," according to Ileen Kemp, the CSLT's secretary, but no statistics were provided for women.[64] In communication with the armed services medical directors in October 1942, the CSLT executive indicated that all the male technicians had likely enlisted.[65] By 1944 nearly ten percent of CSLT members (men and women alike) were in active service for the war effort.[66] The shortage in the armed forces was so acute that the military considered offering short courses for laboratory work, an idea endorsed by the CSLT executive. However, this raised the question what would happen to these men and women when they returned to civilian life.[67] Would they seek employment in laboratories across Canada? While patriotic fervour led the CSLT executive to support any scheme that would ensure an adequate supply of laboratory workers on the home front and overseas, they were quick to defend the drive for standards that was characteristic of the CSLT in its formative years. A streamlined training program, while necessary for the war effort, "may not be ... adequate training for a career as a technician." The CSLT conceded that the training and experience could count toward qualification but asserted that membership in the national society would not be extended to returned men and women automatically.

The question of membership was an important one for the CSLT from its founding. While initially the CSLT granted membership to anyone working in a laboratory, agreeing to "more or less cover all technicians at present employed,"[68] by World War II limits had been placed on eligibility for membership. The CSLT recognized that the war effort demanded quick preparation of technicians for limited duties but understood that the postwar consequences for the professional aspirations of laboratory workers were potentially devastating. In 1941 the CSLT refused several applications from soldiers who were being trained in Toronto.[69] In early 1942, faced with an increasing number of laboratory workers within the military, the CSLT decided to encourage everyone to sit for the national examinations, unless their qualifications were terribly inadequate.[70] In 1944 the CSLT executive decided that "students trained in technical work in the Armed Forces should be given special consideration on their return and some allowance given on their training requirements for the training and experience in the

Armed Forces."[71] There would be no relaxation of the registration requirements for service personnel. The CSLT, having educated hospitals to "insist upon properly trained technicians," did not want its efforts to be undermined by returning men and women with only modest training in laboratory technique.[72]

There were already problems on the homefront. There was a generalized shortage of laboratory workers in the early 1940s as many workers enlisted and as diagnostic services continued to expand. In the midst of labour scarcity, hospitals often found novel solutions. In Montreal, laboratories turned to the Junior League for assistance. Women across Canada were making substantial contributions to the war effort wherever they were needed. The young society women of the Junior League were used to performing volunteer work and may not have objected to working in hospital laboratories for ten dollars a month. But for the CSLT, still struggling with questions of membership and admission to the registry, the presence of the young women and the low salary was an affront to their emerging professional identity.

Although the CSLT found such employment "contrary to the spirit of the newly appointed registry," the Canadian Medical Association did not pursue the matter.[73] Expressing the view of the CMA, Dr G. Harvey Agnew stated that

the presence of these [Junior League] workers, some of whom may not have their honour matriculation and many of whom may not intend to continue the work seriously, does tend to upset the standards of qualification which you have set up but, in matters like this, where long established custom would have to be upset I think that your association would be well advised to make haste slowly. For many years to come some of our larger hospitals will probably continue to train their own technicians, in part at least, without giving much consideration to educational qualifications.[74]

As we saw in the previous chapter, the CMA approved training programs and supported the creation of a registry of laboratory workers. But clearly the concerns and aspirations of laboratory workers were secondary, as Agnew's comments illustrate. The relationship between the CMA and the CSLT was not one between equals.

The shortage of laboratory workers in the 1940s also prompted changes to the registration requirements. Most notably, in the mid–1940s the eligibility age was decreased from twenty-one years to nineteen. This reduction allowed interested students to enter laboratory training sooner than before, which undoubtedly aided recruitment to the bench. This was augmented when the rules were further revised to permit applicants with junior matriculation to be registered. Five years' laboratory experience would stand in the place of graduation from a recognized program until 1 July 1946. Registered nurses would continue to be admitted to the laboratory society without prejudice until that same date.[75] Cumulatively, these changes opened registration to a wider pool of workers. At the same time, however, the changes likely ensured that future laboratory workers would enter training without the benefit of the other kinds of education that were common before World War II, including university and business courses.

It is interesting to see how this contrasts with the drive for membership less than a decade before. Many early CSLT members held only specialty certificates, suggesting that they performed only a very narrow range of tests on a regular basis. In other words, they were not general laboratory workers and not that much different from those who gained training through service in the armed forces. While the early society welcomed such members, the push for standards had made sufficient headway by the mid-1940s that the national office decided to exclude the service personnel.

The shortage of laboratory workers persisted into the mid-1940s. In 1945 the CSLT received sixty-five requests for technicians, of which only thirty were filled through the formal registry. Another twelve were filled through other channels. The next year, there were eighty-three requests for workers, of which twenty-two remained unfilled at the end of the year. The majority of inquiries came from Ontario, which accounted for fifty of the eighty-three requests in 1946, with the rest coming from across the country. Laboratory workers everywhere enjoyed good prospects, however. Sybil Pelton, the secretary of the New Brunswick branch of the CSLT placement bureau, reported receiving requests for technicians from New Brunswick and Nova Scotia but she was unable to find an available technician.[76] Nationally, most of the inquiries came from hospitals, although they also came from other employment streams such as clinics, atomic energy concerns, universities, sanatoriums, doctors' offices, and the Grenfell Association.[77]

As the 1950s dawned, the CSLT struggled to create a stable workforce from which it could grow. In 1951 the society had approximately twelve hundred members, though about ten percent worked in the United States. There were another eight hundred persons working in Canadian laboratories who were not members of the CSLT. In other words, forty-three percent of laboratory workers in Canada were not members of the society.

Significantly, the CSLT estimated in 1951 that there was a shortage of perhaps five hundred technicians and that this deficit would grow more acute.[78] There were many reasons for the continued shortfall. Approximately ten percent of Canadian laboratory workers made their way to the United States each year, to pursue further education or better salary opportunities. The CSLT estimated that another sixty percent of the workers trained each year left the workforce to be married. Finally, in the postwar years, opportunities for women were expanding and applications to laboratory training schools dropped. Only a few years before, school directors could expect as many as 150 applications for fifteen positions. As the 1950s dawned, many classes in Canada were conducted at less than full capacity.[79]

Marriage exacted a heavy toll on the workforce. Several workers from the 1930s and 1940s suggested that married women in the laboratory were unusual, except during the war years. "Perhaps you weren't allowed to work," recalled one worker, adding that at the very least "that would have been frowned upon."[80] A worker from the early 1940s remembers that "when you got married in those days, you left." The same worker also hinted at the class dimensions of such expectations. "I don't suppose anybody was ever married at any job, unless widows, that's about it. Girls from the ... reasonably well off classes didn't expect to work at all ... I always knew I was going to work but not everybody did."[81] Many of the laboratory workers came from the middle classes and they held fast to the notion that married women did not work.[82] The retirements due to marriage, recorded in correspondence and annual reports, corroborate the recollections of workers. There were a few exceptions in Saint John and Halifax, and more during extraordinary periods of labour shortage. As late as 1952, the impending marriage of an employee warranted the comment that the worker was "coming back to work with us after her honeymoon."[83] Resignation from a position in a laboratory usually, of course, meant resignation from the CSLT.

Cynthia Cockburn has observed that "marriage is seen as making a man into a positively better bet as an employee and colleague. It gives him stability, a purpose in sticking to the job. It makes a woman a risk, and childbirth clinches it."[84] Marriage offered advantages to men, while for women the alternatives "too often involved prolonged economic and social dependency or poverty."[85] Nuance is required, however. Marriage also complicated their relationship with the employer for the young men working at Dalhousie. Unwilling to pay a family wage, the university instead opted to employ a succession of young men and bore the cost of the resulting inefficiency and lost productivity. For most women, marriage meant that work at the bench ended. Marriage, then, operated differently for men and women, but it could entail departure from laboratory work for both groups. Men, of course, had the appearance of choice, while women believed, even if erroneously, that they had to leave.

In his 1952 presidential address to the CSLT, Joseph Scott stated that more than two hundred technicians graduated in 1951, but the demand was strong. He continued: "We would like to see more men in the profession, but there is a need for greater economic inducement before we are likely to get them. Salaries in general have continued to improve, but they lag far behind those offered in the United States ... The lure of higher salaries in the United States, and the fact that many girls leave the profession each year for marriage, combine to cause a heavy drain upon the number of available technicians."[86]

Here, in a nutshell, was the CSLT's strategy for professional uplift. It involved expanding the number of personnel and the physical facilities in hospitals so that the laboratory could maintain an efficient service, while concurrently providing an expanded training program to sustain an adequate supply of laboratory workers. Higher wages, ensured through national salary scales, would keep graduates in Canada. Both of these suggestions were grounded in the material conditions that prevailed in the expanding Canadian health care system of the 1950s.

More curious, however, was the suggestion that men should be recruited to the profession. The CSLT executive, prior to the presidential address, identified "the need to improve the economic aspect of laboratory technology in order to offer greater financial security to *men* so that a larger percentage who have interest in the work will find in it the incentive of a decent future."[87] The appeal to male recruitment served to legitimize the need for increased wages for *all* laboratory workers. Concern for prestige or salary are usually reserved for occupations in which men dominate. But in the early 1950s, indeed throughout the entire history of laboratory workers in Canada, women dominated at the bench. In 1951 ninety-six percent of newly registered technicians were women. The same year in the United States, women accounted for fully ninety-five percent of members listed in the American Registry of Medical Technologists.[88] The higher salaries in the United States, often blamed for luring Canadian workers across the border, did little to recruit men to laboratory work.

The ethos of the day suggested that in occupations where women predominate, there was no need to worry about status or providing a living wage. The CSLT knew full well that their American cousins, even with higher wages, failed to attract significant numbers of men. The appeal for more men in laboratory work was a gender-inscribed argument for higher wages for laboratory workers. By articulating a need to raise the profile, status, and wages of the work, ostensibly to attract men, the national society was framing a demand for wages in the language of gender. A "profession" dominated by women is often considered to be somewhat less than a "real" profession; Amitai Etzioni coined the term "semi-profession" to describe such occupational groups.[89] The failure to measure up to idealized masculine notions of "true professionalism" gets translated into lesser status, diminished claims to expertise, and fewer economic rewards. Improvements in these areas could only come through attracting men to the work. When men failed to be recruited in any significant numbers, the claims of women laboratory workers were fatally undermined and the bankruptcy of this approach to professional uplift was revealed.

RECRUITING FOR THE CSLT

Concomitant with the struggle for higher wages and the CSLT's effort to ensure that laboratory workers met minimum educational qualifications, there was the pressing issue of recruiting members to the profession. Even in Saskatchewan, where there were a number of options for those interested in a career in the laboratory, the effort to organize workers into a professional body was stillborn. In 1952 there were eighty-eight hospitals in the province and twenty-one did not have any dedicated laboratory workers. Outside of the larger centres of Regina, Moose Jaw, and Saskatoon, only twenty registered technologists could be found, while thirty-two were working without any kind of certificate. There were also twelve combined x-ray and laboratory workers and another three with "questionable certification."[90] The effort to ensure that laboratories were staffed with workers who were registered with the national society was a dismal failure, particularly poignant in light of the diverse training options open to Saskatchewan residents. While the CSLT and the CMA attempted to nurture a professional identity among laboratory workers, many in the rank and file rejected the effort.

The CMA supported the initiatives of the CSLT to grow. The committee on laboratory technicians reiterated this in the mid-1940s in a report to the CMA executive. The committee felt "very strongly"

that it is advisable to have as many technicians as possible linked up with some official registry such as the CSLT. This is all the more necessary because of the trend of future developments. The want of interest of many technicians in the CSLT is in part due to lack of knowledge of existence of such a body and the advantages to be derived from membership. In this connection the Secretary has written to Miss Kemp [the CSLT president] to give widespread publicity to the present position of affairs.[91]

Isabel Mailhiot, the chair of the CSLT's public relations committee, agreed. For Mailhiot, it was not enough to conduct one's work with "a high degree of technological skill, [and] unswerving moral and intellectual integrity." The laboratory worker must also work at "top efficiency" in order to "sell" laboratory work to other health care workers and the public "as an integral part of the healing arts, as a profession that deserves the respect, admiration and support of society, as something essential to the welfare of the people."[92]

A 1945 *Canadian Nurse* survey illustrates the low profile of laboratory work. *Canadian Nurse* surveyed 566 young women in high-school graduating classes across Canada. Perhaps not surprisingly, fully thirty-four percent selected nursing as their career of choice. More relevant for this study was that the survey offered some data on laboratory work. One Maritimer, three respondents from Ontario and Quebec, two from the Prairies, and one from British Columbia suggested that they would pursue "university, then lab technician." But did this indicate a generalized lack of interest across the nation for laboratory work? The extremely low numbers are likely indicative of the construction of the question. Perhaps had a question been posed about entry directly into laboratory training, increasingly the model advocated by the CSLT and CMA, the numbers would have been higher. "Teaching" for example, received a total of forty answers, whereas "university, then teaching" received only nine.[93] There were other examples. An unspecified booklet outlining hospital careers noted nursing schools across the country, but remained silent on opportunities for laboratory work.[94] If laboratory workers were going to raise their profile among high-school students, they were going to have to do it themselves.

The CSLT embarked on a recruitment campaign in response to the shortage of workers during World War II, the demands of an expanded health care system in the reconstructed Canada, and the short working life of most laboratory workers. The federal Department of Labour suggested that the labour shortage has led to a "vigorous program to recruit and train new workers."[95] The CSLT organized a recruitment committee in 1956 in response to the "acute shortage" of laboratory workers.[96] The committee encouraged members actively to advance interest in laboratory work by speaking to "teenagers or even to those in their twenties."[97] The CSLT regularly distributed career sheets to Canadian high schools as part of its recruitment efforts.[98] Individual members were occasionally asked to speak to groups of interested students. When a Cape Breton worker received such a request in 1958, the national society provided her with a recruitment talk, fifty copies of the "sample career sheet," and two dozen copies of a pictorial pamphlet. Other recommended resources included an American film strip that the CSLT used during the 1950s and a pamphlet from the federal labour department.[99]

The CSLT education committee, chaired by Sister Agnes Gerard of the Halifax Infirmary, completed a pamphlet entitled "If you like Science, why not be a Medical Laboratory Technologist?" This brochure, which was reprinted several times through the 1950s,[100] was aimed at high-school students and provided a general outline of what constituted a career at the bench, answering such questions as "What do you need?" and "Where will you train?" Potential students were warned that "it takes study and constant application" and that this did not end with the training period, because "the good medical technologist keeps abreast of scientific advancement."[101]

In 1959 the CSLT produced another brochure entitled "Medical Technology: A Career With a Future." This pamphlet established the link between "science" and laboratory work more boldly. Laboratory workers were not only key in the fight against diseases such as polio or cancer but were members of "medicine's vast army of professional workers." The authors posed the question how laboratory workers aided the fight against disease and answered in a predictable fashion: the laboratory worker was "a member of a behind-the-scenes corps of workers, a fact finder for the physician," someone who assisted in "carrying on research for new facts and improved techniques."[102] A similar pamphlet produced by the Ontario Hospital Association suggested that young students with an "aptitude for the sciences" could find an "absorbing and satisfying career" at the bench.[103]

The question of salary was addressed in the CSLT pamphlet. It is instructive to see how the national society portrayed itself. "For those university graduates," the brochure declared, "who have post graduate work leading to a master's degree or a doctorate, earnings are the equivalent to those with similar training in other fields. Salaries of from $8000 to $12,000 are being offered microbiologists and biochemists in the very large hospitals."[104] Such extravagant claims were clearly misleading, drawing a spurious connection between laboratory work as a technologist and these other career options. On the surface, they shared a work environment and were joined in their common pursuit of pathogens. But their places in the laboratory and their ability to shape their work were clearly very different. While those holding graduate degrees could expect between eight and twelve thousand dollars per annum, a 1957 CSLT survey found that sixty-one percent of members earned between two thousand and four thousand dollars a year.[105] Those with graduate degrees clearly reaped the reward of their expertise. Laboratory workers did not.

Laboratory work offered opportunities "for both minimally qualified technologists and those holding a university degree." Work at the bench provided an opportunity for those "whose interest is scientific" and offered "a challenge in a field of research in which the enquiring mind need feel no limit." But clearly there were limits. The work was routine, with the same tests being completed day after day, albeit with precision. Laboratory work was not science but "an opportunity for service to humanity." And, of course, there was the wide variety of tasks in smaller labs, where workers should have "knowledge of typing, bookkeeping and filing."[106]

It was, as suggested previously, an area of work that provided opportunity, however constrained, for the women who laboured in the laboratory. The geographic mobility and the opportunity to shape one's working life to suit particular interests were key features of laboratory work in the 1930s and 1940s. It was also a sphere in which married women could work on either a full- or part-time basis, particularly during times of labour scarcity.[107] Workers could seek opportunities in hospital or public health labs, commercial facilities, or industrial, government, or university research labs. A variety of other opportunities were also detailed:

With additional training they may go into other hospital work – medicine, nursing, x-ray technology – to name only a few fields.

University-trained technologists who have a teacher's certificate may turn to the teaching of science subjects in high school or university.

Technologists who have a flair for salesmanship may become sales representatives for large drug houses and suppliers of laboratory equipment because of their familiarity with the needs of the trade.[108]

These, of course, were not really career opportunities for those emerging from training programs. Instead, they all required further training.

DEFINING LABORATORY WORKERS

There is little doubt that the laboratory was an important part of the modern hospital by the mid–1920s, though the development of the labour force was hardly uniform. The laboratory service had proved useful in public health campaigns such as the efforts to provide pure milk or control VD. Laboratories also demonstrated their clinical utility for diagnosing and managing diseases such as diabetes. They were significant sites for both medical education and research. Despite the growing importance of the laboratory on many fronts within modern health care, workers at the bench remained in a precarious professional position, subordinate to the interests of organized medicine and of hospital administrators. The inability of laboratory workers to clearly define the boundaries and to articulate a claim to an exclusive body of knowledge furthered their subordination. Laboratory workers were often expected to perform a variety of duties throughout the hospital. At the same time, other hospital workers would occasionally assume duties in the laboratory. That nurses, dietitians, laboratory workers, and others came to perform a range of duties undermined their claims to expertise. Ultimately, this diminished claim even became internalized into the professional portrait of the laboratory technician.

Evidence for this diminished claim to expertise is disclosed in the records of the CSLT. It is found in their relations with the CMA and within their organization. It extends to their very portrayal of the ideal laboratory technician. In a promotional brochure, the CSLT's education committee put forth the following definition of a medical laboratory technologist as "a person who performs tests in a hospital or medical laboratory; analyzes blood, spinal fluid, sputum, urine, and body tissues in quest of abnormal chemical levels, cells or bacteria; prepares tissue for microscopic examination by pathologist; performs animal inoculations; prepares vaccines; types blood for transfusions; may engage in research."[109] This definition is clearly centred on technique, not knowledge. Technical skill is typically defined as being outside of professional skill; professional skill hinges not on manual dexterity but rather some notion of intellectual mastery. Professionals usually portray themselves as offering a "trained mind," whereas technical workers support the work of others.[110] Despite the efforts of the CSLT to enhance the learned competencies of laboratory workers through new educational and registration requirements, the technique-centred definition of laboratory work endured and continued to ground discussions of skill in the job, in contrast to "skill in the worker."[111]

The emerging portrait of the laboratory worker contained contradictory elements of skill and service that raise questions about historical understandings of occupational groups aspiring for professional recognition. Laboratory work required careful observation and attention to detail, earned through experience and expertise. But while the work required skill, it was also repetitive. Many laboratory workers spent time cleaning dirty glassware, while others routinely filed or typed. Thus, while the actual laboratory analyses unquestionably required skill, many of the other tasks routinely executed by the workers were more humble. The laboratory worker's role in the hospital, therefore, was ambiguous. But such contradictions between skilled work and menial tasks were familiar to many women who worked in health care.[112] Such women shared a bond, insofar as the broad nature of their work resulted in a diminished claim to expertise and to status. The social relations of hospital work, with the multiple demands placed on women workers (and the few men) in the lab, replicated the social relations of other kinds of women's work.

The skilled nature of *some* of the work of the laboratory did not exempt the workers from a wide range of other tasks. The contradictory role of laboratory workers, with their regular multitasking throughout the hospital and their responsibility for both complex analyses and humble tasks within the laboratory, undermined their position within the health care hierarchy. Although laboratory workers possessed the knowledge required to perform their jobs, decisions about which tests would be utilized or what equipment would be purchased were made elsewhere. Nor did laboratory workers determine how the results of their work would be used, either clinically or publicly, or how it was used to bolster arguments in society at large (such as for the drives against venereal disease or polio). Although workers exercised a degree of control over their work, they existed within a bureaucratic organization and were dependent upon that organization to perform their job.

Laboratory workers share with selected other occupations, such as x-ray workers, a history of institutional capture by medical specialties such as hematology, pathology, or radiology. Such capture means that they were continually subjected to a form of medical domination that is perhaps more pronounced than for other hospital workers. Technical workers, as David Coburn et al. have argued, "were 'born' under medical control."[113] Medical dominance was a key feature of their place within the hospital hierarchy. Professional medicine also played a direct role in shaping the professional ideal among laboratory workers through their participation in accrediting training programs, endorsing the registry, administering examinations, and other activities. The professional identity of laboratory workers was closely linked to professional medicine.

In this way, the experience of laboratory workers fits with studies that focus on the *inclusion* of women workers in support of the professional aims of others, rather than those analyses that focus on the history of resistance and professional exclusion that typifies explorations of women in professional work. Tracey Adams has convincingly illustrated that hiring dental auxiliary workers aided the efforts of dentists, in part because it permitted dentists to distance themselves from "lower-status aspects of their work."[114] Like the dental assistants explored by Adams, laboratory workers and other allied health professionals were "channeled into support occupations" that ultimately supported the professional project of male physicians specifically, organized medicine generally, and the labour needs of an expanding health care system in Canada.[115]

It is clear that women were critical to laboratory labour. The analysis of laboratory work in the Maritimes and the national patterns that it illustrates also reveal the multiple roles of women in hospitals. Women's labour was, to put it bluntly, a significant factor in the making of the modern hospital. At the same time, however, their position as interchangeable women workers, capable of filling in when necessary (and in many settings, it was often necessary), limited their claims to full status as health care professionals and ensured their subordinate status to medicine.

"Profession" is a term that is both bountiful and barren for most historians. Its meaning is elusive and, as a category of analysis, perhaps too elusive to be useful.[116] Gerald Geison observed that since publication in 1915 of Abraham Flexner's essay entitled "Is Social Work a Profession?" commentators have struggled with the meaning of the word.[117] Writers such as Donald Scott have argued convincingly that professions need to be understood in their sociohistorical context.[118] Scott's study of public lecturing, which enjoyed a brief period of popularity in the mid-nineteenth century, suggests that professions change over time according to the demands of the public and changing ideas about particular roles and responsibilities. To be meaningful, studies of occupational groups must move beyond internal analyses and question the pronouncements that emanate from the national and local offices of professional societies.

The history of professionals has been dominated largely by studies of particular occupations, with physicians and lawyers perhaps being the best known.[119] The 1960s and 1970s saw a remarkable effort to understand the relationship between social order and professions, an effort that dissipated somewhat in subsequent decades, partly because such analyses are frequently limited by a narrow focus and, more importantly, by a conceptual approach that differentiates them from analyses of other kinds of workers. There are several distinct bodies of scholarship that examine workers and their jobs. The first has class as its central focus, while the second is concerned with "occupations and professions." These two groups of literature are largely separate. Moreover, while they may share an interest in the ways jurisdiction, knowledge, and professional authority are contested, they often ignore the gendered nature of the contests.[120]

The study of "professional women" in North America grew substantially through the 1980s and 1990s and, to a certain extent, bridged this divide because of the uneasy relationship between such categories of analysis as woman, gender, professional, and worker. Many scholars have used the methods of social, labour, and women's history to explore the experience of teachers,[121] nurses,[122] and of female physicians.[123] In

Canada, studies of "allied health care workers," an important employer of women for much of the twentieth century, have begun to appear, including studies of physiotherapists,[124] occupational therapists,[125] and dieticians.[126] Through their analyses of individual groups, these studies reveal patterns that were shared among women. Low status and the accompanying struggle to have their work recognized and valued were familiar battles as women shaped the boundaries of their work. This is brought into even sharper focus in the several books that have explored women's professional work through comparative perspectives, including Mary Kinnear's exemplary study of Manitoba.[127] Interdisciplinarity, well suited to the study of professional work, has been a more recent development. In Canada, the approach is best represented through the research in *Challenging Professions*, an edited collection that offers perspectives on a number of professional careers.[128]

This body of research highlights how women from a number of occupational groups organized themselves into national societies during the opening decades of the twentieth century. Laboratory workers followed this pattern. The Pathological and Bacteriological Laboratory Assistants Association was founded in Britain in 1912, while the American Society for Medical Technologists was formed in 1933.[129] Such organizations, as Nancy Cott suggested, existed for the benefit of the "profession" and actively discouraged "sex-based loyalty."[130] The result was organizations that were constituted by women but that did not exist *for* them. To be a woman professional was to be a person in conflict, for the model professional was assumed to be a man.[131] The CSLT's concern with recruiting men, a concern articulated by other health care professions, affirms this tension between gender and professional identity. Many women recognized the ambiguity and rejected membership in professional bodies. Cott cites the example of the American Medical Women's Association, which never accounted for more than one-third of women physicians in the United States.[132]

The growth of the laboratory workforce, while part of a broader history of the organization of professional groups, also coincided with the rise of white-collar work and the expansion of the public service in Canada. To illustrate, Gregory Kealey has calculated that for every one hundred production workers in 1911 there were 8.6 white-collar workers, a ratio that grew to 16.9 per hundred by 1931.[133] In her analysis of two Ontario factory towns, Joy Parr has also described this transition, as furniture factories began to employ women as clerks and stenographers.[134] So-called "white-collar work" now dominates the landscape of work in Canada. Almost three out of four people in the labour force work in white-collar jobs, and in the postmodern, postindustrial age, this proportion may very well continue to expand. What is important to recognize is that many of these workers, especially those occupying the lower levels of the "white-collar" stratum (in terms of their education, salary, or other such criteria), are subjected to processes of managerial control equivalent to those of the routinized factory worker.

White-collar workers, to which laboratory workers certainly belong, are often viewed as part of the new middle class or as service workers in the postindustrial economy. Others consider such workers to have a contradictory or ambiguous class location, informed not only by a process of proletarianization but also by ideological nonwork criteria.[135] Nevertheless, such workers have largely escaped the attention of historians interested in class, until very recently. The difficulty is magnified for women working in the new occupations of the twentieth century. Women working in department stores, offices, as nurses, or in other occupations of the new middle-class largely fell outside of the research priorities of labour historians who were still seeking to incorporate working-class women into the historical narrative. The result was an "unnatural dichotomy" that ordered class over gender and failed to understand the many similarities among working women.[136]

Even within studies of "professional women" there is a temptation to overemphasize the differences found among occupational groupings, rather than seeking to identify the "common demographic, economic, and cultural conditions to which almost all conformed."[137] For example, in her study of five Canadian professions, Mary Kinnear identified two "governing principles" of being a woman professional: all of the women lacked control over their work and all were paid less than their male counterparts. Yet female professionals enjoyed substantial advantages over their sisters in manufacturing, department stores, clerical work, or personal service. Notably, they earned more money and generally enjoyed greater stability in their jobs.[138] Although there were obstacles, they were further privileged to a large extent by virtue of their education and their race. As Kathryn McPherson suggests in her study of nurses in Canada, the professional bodies and education processes fostered a "racial and cultural exclusivity" that even denied access to some women.[139]

Laboratory work suggests that treating occupations, particularly those that share a work environment such as allied health workers in a hospital, as highly discrete entities is a futile endeavour, obscuring more than it reveals. As has been argued throughout this book, many hospital workers laboured in two services simultaneously and all of them operated under medical authority. All were assuredly subordinated to the interests of professional medicine and frugal hospital administrators. But they also operated within an environment where they interacted with a large number of other workers, similarly organized and sharing similar values. The terrain of health services was dynamic, as new groups of workers carved out niches of activity and old ones redrew or shored up their right to some forms of practice. It is simply not possible to examine laboratory workers or any other allied health group without due attention to the multiple roles that many filled within the expanding Canadian health care complex.

Before 1950 three broad groups of laboratory workers prevailed in Canada.[140] There were nurses who routinely worked in the laboratory, workers who combined duties in the laboratory with work in another service, and dedicated laboratory workers. These classifications were not successive; one did not displace the others. Before 1950 one could find any of these workers in a Canadian hospital, depending on size, location, and clientele. In Halifax, for example, one could find one of the earliest recognized training programs for laboratory workers at the Pathological Institute, while at the Tuberculosis Hospital, which was practically next door, there was a combined laboratory and x-ray technician.[141] Employers searched for workers who could fill either one particular task or, more commonly before 1950, a variety of tasks within the hospital. This was also true of other hospital workers, as the experience of nurses during the 1920s and 1930s illustrated.

The laboratory labour force in Canada shared neither a common education nor a common labour process. Some workers were interested in tissue work, others in serology or chemistry. Laboratory work was not the pursuit of one single science, moreover, but rather a combination of sciences. The particular skills involved in the different departments were quite different, although there were shared core tasks regardless of the setting. The CSLT defined medical laboratory technologists as fact-finders for physicians, the "successors to the work of Pasteur, Koch and others who first brought the techniques of modern science to the practice of medicine."[142] A loose definition to say the least.

Despite the varied tasks, it is possible to reconstruct the idealized laboratory worker and discern what physicians and professionalizers deemed to be the essential qualities for the job. Most important was the emphasis on service. This was an essential component of professionals' self-definition, but there it usually meant service to the public. For laboratory workers, the stress on service was increasingly conceptualized along the lines of hospital efficiency and understanding their place within an emerging complex of service and authority.

The ethos of service, a central underpinning of the CSLT's relationship with professional medicine and hospital administrators alike, was enshrined in a code of ethics presented at the June 1950 annual general meeting. The CSLT developed the code as an "outgrowth of a desire to maintain the dignity and the high esteem of the profession of medical technology. It is a guide for the technologist in all professional activities and relationships." The code's introduction framed medical laboratory technology as "one of the newer branches of the medical arts and sciences" but one with a "worthy role to fulfill." It continued: "The medical technologist, appreciative of the valuable work done by doctors, nurses and others, should endeavour to co-operate fully with them in the care and healing of the sick." This code was reaffirmed in the pledge for medical technologists, adopted at the 1954 annual meeting, which was subsequently recited at graduation ceremonies across Canada.[143]

The idealized laboratory worker was both accurate and honest and reported laboratory findings without suggesting a diagnosis. Laboratory workers were constantly reminded that their role did not permit them to be diagnosticians, effectively separating the manual portion of the work from the intellectual enterprise of drawing a conclusion based on the results. Frank J. Elliott, the first president of the CSLT, commented on this issue at the inaugural annual general meeting, carefully noting that technicians were not qualified to make diagnoses, although they might do so in their own mind. Rather, workers made reports and forwarded them to doctors. Such reassurance was not mere rhetoric and likely came in response to the real fears of physicians. In Massachusetts, for example, the State Society of Technicians was trying to legalize the diagnostic role of technicians.[144] A CSLT pamphlet from the 1950s continued to make the careful distinction between technical work and rendering diagnoses. "The technologist never makes a diagnosis but simply states clearly and definitely what the laboratory findings are."[145] Laboratory workers were to be the technical hands of physicians, at least formally.

The pamphlet, first prepared by two prominent members of the CSLT in 1954, further developed the key attributes of the idealized laboratory worker: "Medical laboratory technology calls for *accuracy* and *honesty* ... the medical technologists must at all times be *exact* and *trustworthy*. An *inquisitive mind*, and an *interest in scientific work* are next in importance. *Manual dexterity* is an essential, but it can be acquired. To have 'good hands' in a laboratory is simply to find oneself at ease in handling the many and varied pieces of equipment. *Co-operation* is exceedingly necessary."[146] Whether working alone in a small hospital or in a complex teaching hospital, the laboratory worker was encouraged to have "good personal relationships" with other departments, doctors, nurses, and all the other staff. Harvey Hall, writing in an early edition of the *Canadian Journal of Medical Technology*, suggested that workers were "not to play favourites. Just because you get along better with one or two physicians does not give you any excuse for doing their work first."[147] Professionalizers like Hall were concerned that social relations would shape the work of the laboratory.

While Hall's remark was not gender specific, it can be read as part of a longstanding concern with women's behaviour in hospital settings. An article published in 1939 made the connection more clearly, declaring that "the young woman whose main object in life is to secure a husband is out of place in the profession."[148] Such descriptions are significant, insofar as they encode stories of skill and appropriate behaviour in the language of gender and ensure that women laboratory workers were seen first and foremost as *women*.[149] The potential for social contact between men and women was great, both in relationships between health care providers (nurse-physician, for example) and between providers and patients. In the early part of the twentieth century, nursing supervisors implemented strict codes of behaviour and dress to govern such social contact.[150] Following the Second World War, concerns about appropriate behaviour in nursing endured. Indeed, the period was a time of heightened heterosexuality, reinforced by images in the popular press and on television.[151]

Concern with social relations among hospital workers found creative expression in a morality play entitled "Team Work – A Time Saver." The play links social relations among hospital occupational groupings with the issue of laboratory (hence hospital) efficiency.[152] Another sketch presented the next year involved the selection of a technologist for a hospital opening. It presents a number of candidates, including Toots Timewaster, Ida Dunno, Miss Intelligentia, and Maud Dell. Needless to say, Miss Dell received the appointment in this farce.[153]

The emphasis on regulating behaviour and ensuring that women in the laboratory behaved appropriately reinforced not only gender norms but also carefully positioned laboratory workers in the hospital hierarchy. Clearly the CSLT, both through pronouncements from Hamilton and through the national journal, emphasized that laboratory workers were part of a health care team. Following this, the emphasis was placed on cooperation with other services and a detached demeanour when organizing one's workday. In other words, personal relations were not to determine when a test was completed. The emphasis on cooperation and harmonious relations was affirmed both by the code of ethics and the graduation pledge. At a time when new workers were assuming duties in the hospital at a remarkable rate and the relations of work *within* the hospital complex were very much in the midst of negotiation, a professional ideology that stressed cooperation served to ameliorate the distinctions among laboratory workers (all workers, regardless of their education or training, adhered to professional strictures) and reinforce their place among health care workers (particularly their subordination to medicine).

Laboratory workers were "only a link in the chain" and "responsibility for action will always rest with the members of the medical profession."[154] Physicians were responsible for the intellectual work of reading laboratory results and making diagnoses or case-management decisions on the basis of those reports. While many laboratory workers enjoyed a variety of roles within the hospital and combined modalities of work, their duties, rights, and responsibilities in all these areas were highly circumscribed. Their place in the hospital hierarchy was entrenched and their working conditions fairly rigid, despite the illusion of option. Not surprisingly, such an environment did little to nurture a collective response to their subordination to physicians or even other health care workers.

The emphasis on providing good service and the regulation of personal behaviour was extended in interesting ways. Much stress was placed on "thoroughness," in addition to the related issue of dedication, with one employer commenting that "I have no use at all for the type of assistant who works with his eye on the clock. The laboratory technician must always be prepared to go on till the work is done."[155] Another author suggested that "of course, we all like our pay cheques. But the true technician is interested in his work to the extent that the money side of it is of secondary importance. If you haven't a genuine love for the work itself you aren't living up to your true capacity as a technician."[156] Laboratory workers were not supposed to accrue material benefits for working overtime, and they were not supposed to be ill either. A writer in the *Canadian Journal of Medical Technology* suggested in 1939 that even a one-day absence disrupted the work of the laboratory, perhaps recognizing the inadequate staffing levels prevailing in many laboratories at that time. The author's remedy was to direct employees "not to go too often to the movies and in times of epidemics of influenza they should avoid them altogether." Workers were to avoid trolleys for similar reasons. To preserve one's health and ensure reliable service to one's employer, "take plenty of exercise in the fresh air both summer and winter and be wise in your dieting."[157]

"Skill" was another important issue, in this case referring to manual dexterity. Technicians were to be "neat handed," and many thought women to be particularly suited to some tasks. Superior workers, however, came in all body shapes. The author of one article cited the example of a "massive bombardier" who could mount incomparable pathological specimens, while another reminisced fondly about an Edinburgh "hunchback" who could produce superior microscopic slides of tissue samples.[158] Manual dexterity could not be guaranteed through a university education, noted one writer, and only natural ability or experience could bestow this quality.[159] A worker from Nova Scotia recalled how excited laboratory director D.J. MacKenzie was when he recruited a young woman from St Francis Xavier University in Antigonish. "She knew all about bacteriology," recalled the worker. "She came, she might have known her stuff, but she was totally useless. She would get mouthfulls [sic] of stuff, and spill things."[160] Such a memory illustrates the ongoing importance of apprenticeship culture that prevailed at the bench, where the content and experience of work were more significant than formal education. What one could do, and how one did it, trumped what one was presumed to know.

Overwhelmingly women, laboratory workers also had to endure an intense personal scrutiny that male workers usually avoided. Registration files routinely included comments about personal qualities. The CSLT developed a form to enable teachers, employers, or supervisors to comment on a variety of the candidate's attributes, including moral integrity, intelligence, dependability, accuracy, cooperation, ability as a laboratory technician, and personal appearance. The form letter also suggested that the information "will be treated with the strictest confidence" and few of the forms have survived. One Catholic sister wrote of another that "Sister appears a little retiring and of a ruddy complexion, average build. To a stranger, she would appear [pr]obably a little stern, which may be due to a little shyness." Nevertheless, the sister thought that the applicant was of "very fine character, serious and very conscientious. She is meek, affable, affectionate and well liked by her clients. Her staff are very fond of her."[161] The same sister submitted comments on another applicant: "Sister has a very pleasing personality, fine looking, rather corpulent and well built. Her appearance is inviting and she should be well received by patients." Besides this inordinate focus on appearance, the senior sister also suggested that the applicant was "confident of her knowledge and is never hesitant in expressing her opinion. She is very alert and bright and she is well trained, in my opinion."[162]

A senior physician from New Brunswick suggested that one of the workers in his laboratory was "very agreeable in the Laboratory and a capable and steady worker. She is well liked. I should think she would make a very good technician." A letter of support from an Acadia University professor suggested that a worker "is neat in her appearance. She is quiet. She has poise. As far as I know her character is above reproach ... She seemed to get along nicely with her fellow students. She cooperated satisfactorily with me. I chose her as my assistant because I considered her the most promising girl in her class."[163] Ralph Smith, of the Pathological Institute, wrote in 1940 that an applicant had attended classes regularly, was a hard and thorough worker, and had "a pleasing tidy personality." Smith also suggested that "his" technicians thought very highly of her: "we feel that she is worthy of being a member."[164]

Writers in the *Canadian Journal of Medical Technology* mocked such personal assessments of applicants' character and the more general efforts to regulate the behaviour of laboratory workers. One article published in 1946 skewered strictures on the behaviour and demeanour of laboratory workers. Margaret Gleason of the Owen Sound General and Marine Hospital described how patients invariably called technicians bloodsuckers or vampires, asked whether they had run out of nail polish (undoubtedly not said to males) or whether they were thirsty, and requested jokingly that not more than a gallon be removed. These queries, hardly original, elicited one of several responses according to Gleason: "(1) Pretend to be deaf, (2) Reciprocate with frayed stock replies. (3) Miss the vein the first three times."[165]

UNIONIZATION

Largely excluded from the emerging portrait of the professional laboratory worker was the unionized laboratory worker. Unionization was not a large issue among laboratory workers, or most Canadian health care workers, before 1950. In 1944 workers at the Regina General Hospital in Saskatchewan successfully organized and became members of the local civic employees' union. Regina laboratory workers were drawn into the labour movement and turned to the CSLT for advice. The national executive discussed the effect of unions on hospitals and Frank Elliott argued that "they go into this union one hundred percent." Although no formal decision is recorded on the matter, Elliott believed that if it enabled the workers to deal more effectively with local authorities, unionization was a good idea. There was another opinion. An executive member from Toronto stated that laboratory workers in that city had recently held a meeting and debated "whether they wanted to be unionized or remain a professional group and they voted 99 percent to remain a professional group."[166] The polarization of "professional" versus union membership was a familiar one among health care workers and the dichotomy engendered many debates within the CSLT.

In late 1948 the CSLT executive took a conciliatory position when another Regina laboratory worker asked about unionization. The executive considered membership in civic unions to be a local matter but thought it better to avoid association with unions "in order to maintain a professional status." Nevertheless, many laboratory workers were civic workers and therefore entitled to membership in municipal unions. The CSLT acknowledged that it "could not prohibit such an affiliation" as long as there was no infringement on the code of ethics and no formal connection between a union movement and the society.[167]

At a western business meeting of the CSLT on the 19 May 1949, members discussed unionization. The chair of the standing committee for the British Columbia branch expressed his opinion that laboratory workers should not join trade unions, which he considered "non-professional." Preferring the status quo, CSLT president Ileen Kemp suggested that the national society neither support nor discourage unionization efforts, although she conceded that in hospitals across Canada, "trade unions have provided material advantages."[168] Both believed that a better answer to labour issues rested with provincial legislation ensuring that only registered workers be employed in hospitals. The CSLT had a decidedly ambiguous attitude toward unionization. Faced with wage disparity across the country, the CSLT believed that mobility of individual workers, not collective action, was the best response. When opportunities were plentiful, many workers protested with their feet, seeking work across Canada or in the United States. When laboratories could no longer retain staff, the thinking went, they would be forced to raise wages.[169] Such a perspective diminished the place of collective action but also suggested that the CSLT would do little to ensure that its members had a strong common voice in labour issues.

The rejection of the labour movement by a proportion of laboratory workers was not unique in health care. While nurses in Quebec ushered in a process of collective bargaining for nurses in 1939 and the Canadian Nurses' Association approved the principle four years later, many nurses continued to believe that membership in unions was unethical.[170] The health care system in Canada at the beginning of the twenty-first century is highly unionized, but most of the organization of hospital workers occurred during the late 1960s and through the 1970s.[171] For the CSLT, inclusion in a union was "detrimental" to professional status. Employers were "antagonized" by unions and laboratory workers who became members, by choice or otherwise, were "naturally tarred with the same stick as everybody else." If laboratory workers wanted to achieve professional status within health care, the CSLT believed they had to resist the siren call of the labour movement. In 1949 the executive stated the matter bluntly. The CSLT should not "affiliate with any union, since that might tend to classify us as labour rather than as professional."[172] The national society resisted any formal relationship with organized labour, though it did adopt the position that individual members or groups of laboratory workers were free to exercise their democratic right to unionize. Laboratory workers, the CSLT believed, were "falling blindly into situations which may be to their eventual detriment if not to their immediate detriment."[173] The CSLT wanted its membership to see the light of professional organization and reject labour organizations.

At the annual general meeting in 1955, the CSLT again addressed the question of unionization. Laboratories in hospitals were being unionized at an increasing rate, "either voluntarily or without knowledge," according to the minutes. "In spite of the apparent advantages of union affiliation," the notes continue, "experience has shown that professional groups have been confronted with serious problems arising from such affiliations. Members of such groups are influenced by union thinking, which in many respects is against our code of ethics." The CSLT suggested that problems in "medical service groups" such as laboratory workers were "vastly different from problems of industry and non medical groups within the hospital." By its own estimation, the national society figured that "non medical" groups constituted over fifty percent of the labour force in a typical Canadian hospital.[174] In defining itself as a professional body, the national society differentiated its members from many hospital co-workers and distanced them from the common struggles that might arise in a shared work environment.

The CSLT then took a much bolder position than earlier. It did not oppose unionization but did "discourage the affiliation of its members with them." The minutes go on:

The individual medical laboratory technologist has a responsibility to the patient and to the service, and that responsibility is realized only by an association such as your representative body. Technologists as members of a service group must be prepared to render service according to need and sometimes beyond regular hours of duty. Their duties should not be subject to the authority or control of bodies outside the profession who do not understand medical laboratory technology, and whose aims and rules may be in conflict with the professional loyality [sic] of the individual technologist. Labour unions cannot offer to the technologist the understanding and leadership that they have in their own Society. We therefore believe that union affiliation should not be sought by medical laboratory technologists for the purpose of collective bargaining.[175]

Janet Plater, the chair of the employment relations committee, then put forward a policy statement that reiterated this position, a position that emphasized service over benefits for the workers, and the professional organization over collective bargaining. There was, remarkably, no discussion from the floor and the statement was adopted unanimously. The ascendancy of the professional ideal over union-based organization and the commitment to service over that of class was complete.

The laboratory worker, in the CSLT's view, should aspire to be an allied health care professional, and the alliance should not be troubled by the politics of class.

There were, of course, other forms of collective activity that flourished in laboratories everywhere. It is not surprising that the laboratory presented an opportunity for women to find friendship with one another. By the 1940s, women in Nova Scotia's public health laboratory worked from nine to five through the week but enjoyed an hour and a half at noon. One worker recalled that the long lunch was wonderful and that she and her colleagues would dash to the Waegwoltic athletic club for a lunchtime swim. Alternatively three or four workers "would race downtown and try to get some nylons" when they were available.[176] After work hours, friends from the lab would "hop on the streetcar and whip down to the Capitol or the Orpheus [theatres] ... we went to an awful lot of shows. And we had parties and all that."[177] Interaction between the workers in the two services, pathology and public health, may have been limited in terms of work, but there was a similar pattern of sociability after working hours.[178] Nicknames were an important part of this culture. Persons separated by such things as geography and age would often refer to one another with a familiar name. Miss Boutlier of Cape Breton became "Boots" to her confidantes in Ontario, while members of the CSLT executive were known across Canada as "Kempie," "Smitty," or even "old girl."[179]

CONCLUSION

It is difficult to gain insight into the day-to-day social relations of the laboratory. Evidence about interaction among laboratory workers or between them and others is scant at the level of the bench or in the individual laboratory. Yet analyzing questions pertaining to employment opportunities and salaries, recruitment initiatives within the CSLT, and the emerging definition of laboratory workers provides valuable insight into the ways in which laboratory workers were incorporated into the health care division of labour and how, in turn, gender was central to that process. In this way, the analysis herein concurs with the views of Celia Davies, who argues that the "central issue ... turns not so much on the *exclusion* of women, but on [the] particular form of their *inclusion*."[180] The example of Esther Stevenson, hired to work in the Prince Edward Island public health laboratory, illustrates the gendered ways in which laboratory workers were added to the health care complex.

Stevenson was a nurse, like many of the early laboratory workers. She therefore fits within the broader story of women who, through their ability to work in a number of services, supported the development of the modern hospital. Stevenson's experience and training in laboratory technique was modest and she therefore worked for less money than a more fully trained worker. As illustrated by evidence from the Maritimes and the country at large, there were wide variations in the wages paid to hospital workers. Maritime laboratory workers believed that this limited their ability to recruit and retain staff, though PEI officials believed that Stevenson could acquire the necessary skills in fairly short order. Stevenson's appointment also reveals something about the broader issues of recruitment and retention of laboratory workers. Women with a number of backgrounds, including both medicine and nursing, were considered for the position. In part, Stevenson was appointed because she had local connections to Prince Edward Island. Local economies were important in determining who worked in the laboratory and how much those individuals were paid.

Stevenson was to be "her own boss," but, it will be recalled, Wodehouse cautioned against hiring an "overefficient" person. Here we see the contradictory ideas embodied in laboratory work. Laboratory workers in many settings were in charge of the day-to-day operations of the lab. But they also worked throughout the hospital, performing many of the tasks necessary to modern health care. Even within the single setting of the laboratory, individual workers combined complex analyses with mundane duties. Finally, laboratory workers across Canada had to contend with the presence of unregistered workers, students, volunteers, and others who laboured at the bench. The lack of a single-portal entry to practice also meant that the skill of laboratory work came to be defined at the bench, rather than in the work, and because of this laboratory skills were closely linked to manual-technical occupations (which required trained hands) rather than to the professions (which required trained mind). All of this undermined any claim that laboratory workers might make to be full professionals and ensured their position as support workers within the health care hierarchy.

A particular way to perform a task, to organize a work day, or to deal with one's patients or peers could reveal much about where one was trained and whether or not a laboratory worker was attempting to fit the work culture of a particular hospital. While useful for inculcating new workers into the labour process of a particular hospital and building solidarity with one's co-workers, the limits of such an approach become readily apparent. To transcend these limits, a broader vision that articulated the values of the rank-and-file had to emerge from the laboratory workshops. Thus, while health care "professionalism" has often been portrayed as a conservative force on an occupational group, it was not inherently so. The CSLT, for example, did consciously reject a definition of its membership as workers. This limited "professional" identity masked interests shared across occupational boundaries and struggles common to other hospital workers. The CSLT's emphasis on a professional model, rather than exploring unions as an alternative way of organizing, was only one choice. It could have articulated a different vision.

Conclusion:
"Standing on the Shoulders of Giants, Leaves Me Cold"[1]

In April 1950 Albert Deutsch published an article entitled "Menace in the Medical Labs" in *Woman's Home Companion*. The brief article, focusing on America, raised concerns about "substandard" laboratories that were committing "fatal errors." The article began dramatically, noting that "Your health – even life itself – often depends on the skill and care of laboratory technicians serving your doctor or hospital. Many are wonder workers, indeed. But a nation-wide survey reveals a frightening increase in error and carelessness – a betrayal of trust that can kill." This ominous message, which impugned laboratory and education standards, caught the attention of the CSLT executive the same month it appeared.[2]

Despite the sensationalism, the article conceded that most laboratories were "excellent" but noted there was tremendous variation from setting to setting. Laboratories, it reported, varied in size from a single room, "manned by one or two technicians able to perform only the simplest tests," to very large facilities with "scores" of employees performing "the most complicated analyses." The article noted that some laboratories were led by clinical pathologists ("men with medical degrees"), while "biochemists and other scientists" directed other facilities. "Under them [the laboratory directors] are medical technologists with enough training to give them semiprofessional status. At the bottom of the laboratory hierarchy are the technicians, often young girls with little education, hastily trained for routine tasks." Deutsch, citing the Pennsylvania Society of Clinical Pathologists, reported that the chief cause of laboratory error was "poorly trained technicians," although there were other factors, including inadequate equipment, poor facilities, and staff shortages.

The drama of "Menace in the Medical Lab" captures some of the tensions that existed within medical laboratories at mid-century. While much had been accomplished in terms of framing laboratory work as a profession, the claims remained tenuous. The CSLT played an active role in legitimizing a portrayal of laboratory workers that emphasized the divide between manual and mental work and concurrently valorized the latter. Ultimately, the CSLT drew a division between laboratory workers and other hospital workers, creating a social cleavage within the hospital that left laboratory workers on rather tenuous ground, with a highly ambiguous identity. This is all the more surprising given the nature of laboratory work during the 1920s, 1930s, and 1940s.

It is important to acknowledge that laboratory workers were part of a system of professions within health care. But the more important interpretive point is that individual workers were also members of several different occupational communities simultaneously. The study of laboratory workers highlights that they were also nurses, dietitians, x-ray technicians, etc. The implications of these multiple identities for our understanding of health care work are in need of further exploration and I will leave this task to future research projects. What is clear is that the analysis that emerges from a focus on a single group, laboratory workers, carries implications for many other groups.[3]

The history of health care, as many have acknowledged in recent years, is more than the history of physicians or the institutions of health care. More than a decade ago Wendy Mitchinson sensibly entreated historians to engage in understanding the patient experience and scholars have begun to take up her challenge. Nurses and other occupational groups have been explored in different settings and to varying degrees of depth, though the project remains woefully incomplete, given the important role of health care work as an employer for women. The analyses of health in specific ethnocultural communities in Canada have established linkages across historical genres and have skilfully deployed the tools of social history, gender studies, cultural studies, and a range of other analytical approaches.[4] Alison Prentice has argued that "we need women's stories as well as men's if we hope to develop a full picture of how science works."[5]

To this I would add that, if we are to better understand the making of the twentieth-century hospital, we need explorations of women's work in the nooks and crannies of the hospital. We need to consider the activities of technical workers, housekeeping staff, kitchen personnel, secretarial staff, and others. Hospitals are emphatically not merely sites of work for physicians or nurses, although the experience of other workers is difficult to fully reconstruct. The purpose of exploring such workers is that it illuminates the relationships that existed among them and, in turn, how they were integrated into, and challenged, prevailing knowledge systems and ideas of expertise. Analyses of these workers, as the example of laboratory workers vividly illustrates, also reveal much about the social organization of health care work and the significant role of broader economic patterns in shaping hospital work. Indeed, excluding broader social and economic factors from analyses of health care work is sheer folly.

Throughout this book, I have paid attention to the duties of laboratory workers in specific contexts. Many of these workers began their careers as nurses but had stays of varying length at the laboratory bench. Others crossed the often fuzzy boundary between student or volunteer in the laboratory to paid employee. In addition to their laboratory duties, workers often had responsibilities in departments such as nursing, dietetics, x-ray, or pharmacy. There were few sharp distinctions among areas of service in the hospital before 1950. This afforded workers an opportunity to shape their work within the hospital. Workers could advertise their skills in a variety of services and choose among hospital departments. Boundaries were fluid, as hospitals filled openings in new services and workers struggled to meet new demands. The experiences of laboratory and other hospital workers belied an emphasis on unfettered specialization as the twentieth century progressed. Workers did not always perform discrete tasks, which undoubtedly served the interests of hospitals but also provided workers with a degree of mobility and some control, albeit limited, over their worklife.

Understanding laboratory work, and other aspects of health care work, also requires some consideration of local, national, and international developments. What, then, were the important features of laboratory work revealed through the Maritime experience? At the local level, the study of laboratory work reveals the participation of multiple levels of government, often working closely with other agencies or institutions to put in place the infrastructure for a "modern" health care system. The evidence from the Maritimes suggests the uneven development of health services, not just between the Maritimes and other parts of Canada but also between rural and urban settings or large and small hospitals. In this way, the Maritime case study presented herein offers insights into laboratory work that have national implications for our understanding of the history of health services. The case study also challenges the "aura of inevitable permanence" of occupational groups and, in this way, contributes to an international literature on health care work.[6] Perhaps most importantly, the Maritime case study illustrated the multitasking that prevailed in many hospital settings. A second key feature was the multiple paths to the laboratory bench, which included nursing education, university courses, short courses in laboratory technique, and longer training at the bench.

The lack of a common route to the bench well into the 1950s and the multitasking that prevailed permitted lab workers to make only tenuous claims to knowledge and skill. Laboratory workers, in an age when new tests were being added, did not successfully articulate a claim to that body of knowledge. In part, this was because they shared the terrain of the laboratory with other workers, most notably nurses. The flexibility that was a feature of many aspects of hospital work was particularly damaging to an emerging labour force within the hospital. The ambiguity of laboratory roles essentially meant that all laboratory workers were grouped as an undifferentiated mass. There were students, those who were labelled "technicians," or those merely providing laboratory "services." The fact that these individuals could not be distinguished from one another homogenized and effectively devalued laboratory work to a large degree. What is most striking is how this sharply contrasts with the growing specialization of medicine and, concurrently, departments within the hospital. This confirms the insight that women workers, in this case laboratory workers, were viewed as interchangeable, while physicians, most of whom were men, had recognizable and individual skills.[7]

Equally important was that laboratory workers had no exclusive ownership over skills. The debate over skill is, of course, a rich one historiographically.[8] What is most germane to the laboratory is that physicians occupied the critical position of laboratory director and it was they who shared in the sociocultural network of medicine that valorized the interpretation of results over the preparation of those same results. The skill attached to the reading is what was important and shaped how work was organized and rewarded within the hospital. Few women and virtually no laboratory workers shared in this culture. Rather, like many technical workers, those who laboured at the bench occupied an interstitial position, serving as a buffer between the patient and the physician, between medical technology and the public.

The pattern of multitasking among laboratory workers, the specific details of which have been documented in this book, suggest that hospital work was rather more fluid than orthodox accounts permit. For this reason, I wish to argue, in the same vein that Gregory Kealey has mined, that workers such as laboratory technicians are integral to our understanding of the history of health care in the twentieth century. Furthermore, it is my contention that no portrayal of the modern hospital or public health system is complete without some consideration of these workers and the broader issues that they raise for our understanding of the organization of health care work.[9]

There is, then, a need to reorient analytical conceptualizations of hospital work away from discrete units to larger frameworks. It is critical to remember that health care occupations interact with other occupational groups on a daily basis and, in turn, affect and are affected by the nature of those interactions. At the same time, the idiosyncratic nature of hospital work, which varied not only between hospitals but could very well vary between night and day shifts or among floors, suggests the need to understand health care professions in specific work contexts and in terms of their relations with many other health care workers, and not only their interactions with organized medicine. The history of the hospital and of health care labour looks different from the perspective of workers.[10] This examination of laboratory workers has suggested that the hospital is really a series of linked work environments, many of which are gender-bound, that share the same physical space.

It would have been easy to speculate that the Maritimes were unusual because of regional deindustrialization and underdevelopment or because of the many small community hospitals dotting the rural landscape. This book, although firmly situated in the historical experience of the Maritimes, examined developments elsewhere in Canada. The evidence is conclusive; the pattern of multitasking by women laboratory workers identified through the Maritime case study is characteristic of other settings as well. Perhaps there was something unique about laboratory workers that linked their kind of work to others in a way that was not typical of other hospital workers. This does not seem to be the case either. Nurses, dietitians, and x-ray workers were all interconnected as well, with the laboratory and with one another, though we know very little about the meanings of those connections for those workers.

A more orthodox interpretation, such as the institutional framework developed in the first chapter, might have followed the growth of the laboratory service as *part* of an already assumed story of hospital departmentalization and increased specialization. One could point to the foundation of the different kinds of laboratory work, including the development of serology, bacteriology, chemistry, hematology, virology, and pathology, dutifully tracing the origins, adoption, and growth of the various tests used in each group. One could examine the growth of formal training schools with an approved curriculum, the creation of entrance standards to those programs and the institution of a formal examination as evidence of professional formation. One could even assess the relative shortcomings and strengths of the CSLT's drive to organize and recruit members or negotiate and carve a place for those members among other health care workers. What such an interpretation would miss are the very real ways in which laboratory workers contradict this assumed story. Theirs is a story of diverse routes to the laboratory, diversity at the bench (and among CSLT members once it was established) and multitasking. The implications of the organization of work in this way remain unexamined, indeed largely unacknowledged among historians of health.

Gerald Larkin has noted that the integration of allied health care workers into the system that exploits them poses challenges to Marxist interpretations. Some, such as Eliot Freidson, argue that professionalism is utterly bankrupt as an ideology for health care workers because it is stripped of any notion of work autonomy. As Larkin wrote "it is not explained why para-medical workers pursue a 'mirage', or the control of their work within a system based upon its negation."[11] It is not sufficient to explain workers' attraction to professional ideology in terms of a false consciousness. The status of laboratory work, like other kinds of women's work in health care, was both ambiguous and uncertain. Health care was dominated by male physicians and the interest of professional medicine. As an occupational group, laboratory work was divided internally by such elements as educational background, training, and whether employees worked exclusively in the laboratory or simultaneously in several departments, what we might term their practice modalities. Laboratory workers, birthed in part by the needs of organized medicine, never enjoyed the kind of independence usually cited as a characteristic of professions. Working in concert with organized medicine, the CSLT did exercise some influence over the organization of laboratory work and its content. They were also able, by assuming positions as teachers and as senior laboratory staff connected both with physicians and hospital administrators, to shape issues such as the training and certification of laboratory labour.[12] The CSLT also played a critical role in recruiting workers to the occupation and in defending the occupational boundaries. Nevertheless, laboratory work was part of the machinery of health care and clinical information. By aligning themselves closely with professional medicine, laboratory workers gained a degree of recognition but also created the conditions for their subordination.

Institutional development was an important part of the public health effort in the opening decades of the twentieth century. While the role of the laboratory in public health is most often emphasized in existing studies, and rightly so, our view of the laboratory should not be limited to public health work. In Saint John and Halifax, laboratories conducted clinical work for hospitals throughout the Maritimes, thereby consolidating the good will of practitioners beyond the city limits. In New Brunswick, the laboratory was conceptualized from the beginning as an integral part of a robust department of health, but its influence extended even into Prince Edward Island. Most explicit in Nova Scotia, the Morris Street laboratory was a place of medical education and the new facility constructed in the wake of the Flexner Report did much to "modernize" instruction in selected medical sciences. The facilities in Halifax and Saint John were never narrowly conceived.

While the institutional history of the laboratory may seem insignificant, it reveals how multiple interests shaped the work of the laboratory, including different levels of government, hospitals of varying sizes, and the medical school. This, in turn, shaped a perception of workers as flexible labourers who should be able to undertake a broad range of work. In smaller community hospitals, this extended well beyond laboratory work to other departments altogether. The principal laboratories in the Maritimes were not important institutions. They did not make any great bacteriological discoveries or significantly advance medical science. The laboratories in Saint John and Halifax, however, stood as a clear manifestation of the scientific and diagnostic power of the new public health. In this way, the local developments clearly reflected international developments in science and medicine.

By the middle 1890s many jurisdictions in the United States and Canada had established diagnostic laboratories and within a few years they had become an integral part of any modern health department. Thus, in 1918 when New Brunswick set out to establish its new department, complete with the first minister of health in the Dominion of Canada, a laboratory was an integral part of the plan. Routine testing became common and there was a movement toward extensive testing for a host of diseases. In John Duffy's words, "science rather than sanitation now seemed to be the solution to sickness and disease."[13] The creation of laboratories, by providing vaccines or antitoxins or revealing vectors and healthy carriers through lab analyses, also solidified the tentative link between science and medicine.

But this is not the story of competition between those who advocated "scientific medicine" and those who maintained their faith in the clinical examination. Indeed, the establishment of facilities in Saint John and Halifax reveal the cooperation between community physicians and their brethren in the public health departments, the medical school or the laboratory itself. Essentially, there was no sharp division between the two in the Maritime context. This was not merely good fortune. The laboratories in Halifax and Saint John were part of the new health consciousness that manifested itself in such diverse activities as medical school reform, the activities of the Massachusetts-Halifax Health Commission, the renovation and expansion of large hospitals in Halifax and Saint John, and the construction of community hospitals throughout the region. The laboratory stood as testimony to the commitment of various provincial governments to improving the health conditions of the population and was part of a larger reform movement. Yet while provincial governments were willing to support the establishment of laboratories, there were significant constraints. As the optimism of the 1910s evaporated in the wake of the economic collapse of the Maritimes during the 1920s, complaints about inadequate space, faulty or missing equipment, and frequent laments about inadequate staff were regular mantras in the laboratory reports. What the complaints reveal are the very real limitations on public health in the Maritimes that were apparent by the 1920s.

Nevertheless, the laboratory was clearly ascendant in the Maritimes on a variety of fronts, including medical education, public health, and clinical care. The close relationship between the medical profession and a variety of private and state interests ensured this ascendancy. The same relationships shaped the work of the laboratory. This is most apparent in Halifax, where the province, city, and university all contributed to the operation of the laboratory. The Pathological Institute also assured residents of Nova Scotia that they had potable water supplies and that milk producers adhered to some level of sanitary standard. Laboratory analyses revealed threats to the public's health, from diphtheria to syphilis. The institute also made health a priority for everyone, since an apparently healthy individual or "pure" water supply could infect unsuspecting individuals. Equally significant in shaping the work of the laboratory was the burden of disease. The utility of laboratory work was clearly demonstrated in the campaign against diphtheria and in the diagnosis of tuberculosis and venereal diseases. As new diagnostic arrays became available, they were incorporated into the test battery. Municipal and provincial governments alike wanted to ensure that milk and water testing was available. As the work of the laboratory grew and the facilities were established, then expanded, additions were made to the staff. Laboratory workers grew from a handful of individuals in the early 1920s to an important service by the end of the decade, one that encompassed a broad spectrum of work, including serology, bacteriology, and hematology, to name a few.

The local study of laboratories and other services reminds historians that these are social institutions, founded to meet local needs. Hospitals insert themselves into the community in a number of ways. They provided training for several generations of nurses and education for the bench was centred there as well. The first applicants to the informal laboratory courses in Halifax and Saint John were nurses, for whom work at the bench was a regular part of their duties. Training programs were initiated in the 1920s and grew more formal and regular through the 1920s and 1930s. In New Brunswick, one of the functions of the provincial laboratory was to train workers to staff smaller clinical laboratories throughout the province. The demand for training was prompted by the expansion of hospitals and the services they offered, along with external factors, such as the hospital accreditation movement. The result was that new workers were introduced to the hospital. Regardless of whether they came from a nursing background, secondary school, or university, all workers learned at the bench. One learned by doing, and didactic instruction was minimal. The laboratory course in Saint John began in 1919 after Harry Abramson joined the Department of Health and drew students from across New Brunswick, from the southwestern portions of Nova Scotia and from Prince Edward Island. There was no beginning date for courses, but students were enrolled when the laboratory could accommodate them.

The growth of facilities and the expanding workforce must not be assumed to be evidence of the increasing and linear trend toward specialization. To reiterate, the laboratory worker may have performed bench work but often filled a variety of roles within the hospital. Laboratory workers were a diverse lot. The specialization that is so often touted as a characteristic of the modern hospital is severely complicated by the experience of laboratory workers. The creation of a labour force for the laboratory and the other hospital service departments was guided by a number of factors. The new services came under the direction of physicians, while trained workers carried out the routine work. Nurses filled many of the earliest positions in the laboratory, particularly in the smaller hospitals that were being established throughout the region. Nurses were expected to fill any number of roles within the hospital, including the laboratory. Other workers also had the same expectations forced upon them. That is to say, persons working in labs also supplied other departments as well. There was, then, a clear emphasis on fulfilling a variety of roles within the growing hospital complex. This obviously served the ends of hospitals, but it may also have allowed workers to shape their own career. Advertisements in medical journals detailed the skills that individual workers possessed, and these could be manipulated to secure a desired position.

The knowledge and skill set of laboratory workers is an ongoing matter of negotiation. Jeanne Irwin, the president of the CSLT in 1996, acknowledged that the society was entering a period of transition. Diminishing levels of staff or of persons entering laboratory work would certainly take a toll on the national membership. Irwin believed that the national society would have to "encompass laboratory assistants and other lab workers as part of the membership." Persons with baccalaureate degrees, a BSc in microbiology, for example, might work next to an individual with a Registered Technician designation. Laboratories that are more automated may offer less opportunity for "hands-on" work but will demand that workers acquire a "lot of different training and expertise to be of value."[14] The downsizing that was so characteristic of hospital work environments through the late 1980s and 1990s brought history clearly into focus. There was talk of upskilling, multitasking, and flexible-specialization, the buzzwords of an age trying to come to terms with fiscal constraints. Yet these new demands on workers have historical precedents, one of the continuities between the past and the present. What remains to be negotiated is whether laboratory workers can assert some authority and claim economic reward, job security, or enhanced status.

In advance of the diamond jubilee of the Canadian Society of Laboratory Technologists in 1997, the national organization turned an eye toward history. They ran a series of profiles entitled "On the Shoulders of Giants" that purported to recognize individuals for their "extraordinary contributions to the growth and development" of laboratory work in Canada. The CSLT wanted to acknowledge exemplary individuals that aided "medical laboratory technology to reach the high standards of professionalism it enjoys today."[15] One of the subjects of these profiles, Norman Senn, did not apparently consider himself a giant but rather "a minor person in a great organization."[16] What made a "giant" in laboratory work? One could suggest that the most famous laboratory worker ever was Mary Mallon. Mallon, "Typhoid Mary" as history remembers her, began to perform selected bacteriological tests while being held in quarantine.[17] She is probably not the figure one wants to associate with an emerging profession. Of course, there are the great breakthroughs of the laboratory but these hardly belong to the technical hands who for the most part remain obscure, even in the most critical accounts of laboratory discoveries.

Laboratory workers are hidden from view. Samples are sent "down" to the laboratory, regardless of where it is actually situated within the modern health care complex. In the imagination the laboratory is always in the basement, and this is a profound signifier of the fact that laboratories are subordinate to other health services and also invisible to the public. The theme of the 1998 National Medical Laboratory Week was "Reaching beyond Technology to Discover the Secrets of Your Health."[18] The theme was selected in part to "create a personal link between the medical laboratory technologist and the community by emphasizing the important role technologists play in an individual's health care."[19] A recent provincial civil service campaign in Nova Scotia, entitled "The Secret Service," included laboratory workers among many of the invisible workers essential to the operation of the modern hospital. A decade earlier, medical technology week was aptly titled "the quiet perfectionists."[20]

Laboratory workers, in common with health care workers of all kinds, have endured a difficult period during the 1990s, sometimes quietly but often seeking to be heard.[21] Hospitals have been closed or restructured and opportunities for laboratory work have become increasingly constrained. In Nova Scotia, the medical laboratory technologist training program at the Community College's Institute of Technology campus was suspended because of the lack of jobs in the field and, presumably, as a cost cutting measure within the Department of Education.[22] Under the auspices of reform, staff levels have been reduced, laboratory services have been amalgamated or privatized in the pursuit of rationalization, and workers have been asked to take on an increasing level of responsibility for authoring methods or approving new equipment, without corresponding wage increases. The Canadian health care system is under tremendous pressure at the beginning of the twenty-first century, having become subject to a variety of competing claims from "stakeholders," an elusive term that includes patients, health care providers, hospital corporations, and private companies all eager to take over responsibility for portions of health services, including laboratory work. Yet the development of laboratories in the Maritimes and beyond was always dependent upon the confluence of a variety of interests at the bench.

In contrast to the "giants" are the many laboratory workers who continue to see themselves as "just bench techs." But these are the workers who are the vast majority of the membership. Call them the rank and file, as historians are apt to do, or the "heart and soul" as workers prefer; the vast majority of laboratory workers will never occupy a place of national or scientific "significance" as it is usually conceptualized. But they do perform essential tests that benefit patients and doctors. They are an important component of the modern health care system, even if they are a "secret" one. There is no such thing as "just a bench tech." As Ruth Pierce suggested, "I am a medical laboratory technologist. I am employed by the department of Microbiology at the IWK/Grace Health Centre [in Halifax]. The key word in 1996 is employed. To identify myself as 'just a bench tech' I might just as easily say I am 'just a woman' and you'll not hear that."[23]

There is no doubt that gender was a significant operative in the expansion of laboratory work. To pick up on the insights offered by Tracey Adams, laboratory work in health care was not gender-neutral.[24] In many settings, the tight budgets of the 1920s and 1930s and the growing staffing needs of hospitals demanded that workers fill multiple roles and women – already working in many capacities in the hospital – were ideal candidates to assume new duties. Laboratory work was and is conducted in settings with a complex division of labour, shaped not only by health care specialization and changes in technology but also through a range of administrative needs and constraints. Women's laboratory labour was a critical piece of meeting these needs. At the same time, the example of the Dalhousie laboratory boys illustrates that gender did not operate alone or in isolation.

The differently patterned gender relations at Dalhousie and the Pathological Institute suggest the complex interplay of class and gender, as many accounts have acknowledged.[25] But *setting* was also key; the expanding infrastructure of health care and the limited financial resources of the university, both rooted in the broader material conditions of Nova Scotia in the 1920s and 1930s, intersected with both gender and class to structure the laboratory labour force. Joy Parr, in her analysis of two Ontario factory towns, concludes that "while the theoretical literature on gender segregation in the labour force is rich, economists and feminist theorists have been interested in sexual divisions as general features of the economic or sex/gender system rather than as boundaries between tasks forged in defined contexts by particular clashes of interest." Setting, in combination with gender, class, and ethnicity, helped to shape the factory labour forces described by Parr.[26] The interpretive contribution of the study of laboratory work is that it sharpens the importance of place. The many comments about "local girls," also found in Parr's study, and the differing gender profile of two laboratory labour forces in Halifax confirm the importance of setting and the need for specificity.[27]

Again, the local study of laboratories reminds the historian that these are not simply places of science but also sites of work. The technical hands may be unimportant to the person interested in clinical care, but reconceptualizing the laboratory as a place of work opens new lines of inquiry. Much of the debate over salaries within Dalhousie University concerned the question of family maintenance. Men were usually thought to be primarily responsible for sustaining families; even if their children were contributing to the family economy, as in the case of Albert Hallett. Marriage complicated the discourse surrounding wages in the Dalhousie laboratories. Clearly the university, in keeping with its philosophy of never paying a family wage, tried to employ only young men. The tender age of the recruits was reflected in the low wages paid to these workers. The university maintained that the work was not suitable for a married man who had a family to support. Men who worked in laboratories were also thought to be less ambitious than other men. The university believed that determined men would simply look for more remunerative work, despite the obvious recognition that an effective laboratory worker aided instruction and research. The skill of these workers was not acknowledged either ideologically or financially. They could simply be replaced. The university was willing to bear the cost of frequently having to train new workers, rather than paying a wage sufficient to retain competent staff. Age, gender roles, and a belief about ambition in the capitalist world combined to justify the low wages paid to the support staff in the university. What is interesting is that any justification was proffered at all. For the women across Morris Street or in Saint John, no explanation was necessary. After all, the entry of women into paid work was presumed to be temporary and their wages were not necessary for family maintenance. The wages for a single woman such as Margaret Low could be depressed with impunity. Only when her brother, who had provided partially for Margaret's well-being, died did the question of *adequacy* ever emerge, and then only briefly.

Women did not enjoy the same range of opportunities as men. There was a wide range of constraints on the working lives of Maritime women. Even elite women, such as female physicians, faced a dazzling array of obstacles to establishing successful practices. Some were forced to practise in rural or less-desirable locations, while others abandoned medicine altogether for more "suitably female" pursuits. Many of the earliest laboratory workers had the benefit of a good education but despite this their career prospects remained limited. A job in the laboratory offered them a significant alternative to other forms of work. As a friend who works in a laboratory once told me, "Nursing is like being a maid, while laboratory work is like cooking, and I would rather be a cook than a maid." The comment is deeply inscribed by gendered assumptions but working in the laboratory did offer women a chance to do something other than tend to the demands of their own families, care for the sick, or teach children. It also allowed women with an aptitude for science to pursue their interest, although in a highly constrained way. Nevertheless, laboratory workers were also relatively privileged. They worked regular hours, were paid a decent if not exorbitant wage, and were entitled to annual vacations. For some, it was a good job.

At the same time as the CSLT was planning its sixtieth anniversary and honouring its past, it also changed its name. It became the "Canadian Society for Medical Laboratory Science" and the new identity was to give the membership "a new face and a new pride in our profession."[28] The emphasis on science was not new, but the name is certainly more inclusive, intended to expand the membership from technologists to include those with baccalaureate or graduate degrees in the medical sciences, PhD researchers, and even medical specialists such as hematologists. The need to expand the membership of the national society beyond the bench worker is a challenge for the new millennium. In a sense, it marks a return to the origins of the national society.

There was an overwhelming desire to standardize the experience of becoming a laboratory worker. Efforts to establish a curriculum, national examinations, and registration were designed to ensure that laboratory workers beginning their careers at the bench met some minimum standards. At the same time, however, the CSLT was a national society that sought to accommodate diversity within its membership. It was a body steadfastly committed to maintaining "professional" standards and vigorously opposed to definitions that diminished the perceived status that accompanied such standards. The society articulated its independence, yet clearly remained subservient to the interests of the medical community that set its examinations and demanded stringent registration requirements. It struggled to define itself as a national society and confronted the difficulties of all such bodies in maintaining sufficient strength and interest among its members.

In the diversity of those who laboured at the bench laboratory workers faced an additional obstacle to their effort to create a national organization. Nurses, laboratory workers, and combined technicians all found employment in the laboratory, and indeed all were included in the nascent CSLT. Membership, as we have seen, was initially defined not through education or training but through skills. For laboratory workers, a definition based on skills takes on new significance in the midst of decentralized training, varied educational qualifications, and different patterns of work. In order to organize the bench, leaders had to construct common ground and the only way to accomplish that was through a description of characteristics. Precise enough to form common ground but vague enough to be inclusive, the idealized laboratory worker could be a nurse, a university-educated technician, or a hospital-trained laboratory worker. Laboratory workers reveal the limits of approaches that seek to explore only the creation of discrete occupational groups and fail to account for the social relations of work. The complex world of laboratory work renders the search for giants meaningless. The real explanatory power of studying laboratory workers is only revealed through the common experience of the many who were not "just" bench workers but, first and foremost, workers in a complex environment.

Notes

ACKNOWLEDGMENTS

1 CSLT, Minutes of the Executive, 29 March 1953.

PREFACE

1 Kelm, *Colonizing Bodies*, received the Canadian Historical Association's John A. Macdonald Prize as the best book in Canadian history in 1999.
2 Godfrey, *The Struggle to Serve*; MacLean, *Asylum*; MacDonald, *Mount Hope Then and Now*; Howell, *A Century of Care*.
3 Amirault, "The Historical Evolution of Nursing Education"; Twohig, *Challenge and Change*; Callbeck, *A History*; McGee, *The Victoria Public Hospital*.
4 Ian Cameron, "The Quarantine Station," 1938, "Camp Hill"; Murray, "The Visit of Abraham Flexner"; Penney, "Marked for Slaughter"; Keddy, "Private Duty Nursing Days"; Scammell, "A Brief History."
5 Waite, *The Lives of Dalhousie University*, vols. 1 and 2, and James Cameron, *For the People*. For an interesting account of Maritime women who left the region to become or work as nurses, see Beattie, *Obligation and Opportunity*, ch. 4.
6 Baldwin, *She Answered Every Call*, "The Campaign against Odors," "Volunteers in Action."
7 NSARM RG 25, Series C, Vol. 1, Andrew Halliday to A.P. Reid, 29 January 1901.

CHAPTER ONE

1 Cavanagh and Bamford, "Substitution in Nursing Practice," 333–9; and DiCenso, "The Neonatal Nurse Practitioner," 151–5; Burtch, *Trials of Labour*.

2 Gilbert, "Dispensing Doctors," 83–95; and Ritchey and Sommers, "Medical Rationalization," 117–39.
3 Halpern, "Medicalization as Professional Process," 83–95.
4 Begun and Lippincott, "Origins and Resolution of Interoccupational Conflict," 368–86.
5 Biggs, "User Voice," 195–203.
6 Bickerton, "Reforming Health Care Governance."
7 For recent analyses, see Soothill, Mackay, Webb, eds., *Interprofessional Relations*; Owens, Carrier, and Horder, *Interprofessional Issues*; Leathard, *Going Inter-Professional*; Ovretveit, *Co-ordinating Community Care*; Lonsdale, Webb, and Briggs, *Teamwork*.
8 In 2000 nurse practitioners were introduced in four demonstration sites throughout mainland Nova Scotia. Nurse practitioners have been studied extensively. For articles in the primary care setting, the nexus of interaction between nurse practitioners and physicians, see Archangelo, Fitzgerald, Carroll, and Plumb, "Collaborative Care," 103–13; Buchanan, "The Acute Nurse Practitioner," 13–20; Sebas, "Developing a Collaborative Practice Agreement," 49–51. For an article that directly addresses the question of practice scope and professional identity, see Brush and Capezuti, "Professional Autonomy," 265–70.
9 There are many examples in the health service literature that can be categorized. Those focusing on encroachment or competition include the efforts by US pediatricians to expand their practice roles after World War II (Halpern, "Medicalization as Professional Process," 28–42, and *American Pediatrics*), efforts by South African pharmacists to prescribe for patients (Gilbert, "Dispensing Doctors," 83–95), or the long-standing attempt to expand the scope of optometry to include prescribing drugs for the eye, historically limited to ophthalmology (Begun and Lippincott, "Origins and Resolution," 368–86). Efforts to redraw or redefine boundaries often focus on efforts to substitute "unlicensed" or "paraprofessional" personnel for tasks formerly performed by "professionals." The advance of nursing aides is perhaps the most familiar example. For an early analysis, see Strauss, "Structure and Ideology of the Nursing Profession." An example of reprofessionalization may be found in hospitals among clinical pharmacists who are now involved in patient management and treatment decisions, expanding their professional boundaries as members of the "medical team" (Mesler, "Boundary Encroachment," 310–31; Ritchey and Sommers, "Medical Rationalization," 117–39).
10 Fahmy-Eid, *Femmes, santé et professions*.
11 Larkin, *Occupational Monopoly*, vi.
12 Buxton, "Private Wealth and Public Health," 183–93. The explosion in Halifax Harbour occurred on 6 December 1917 and was the largest man-made explosion before Hiroshima. It occurred when the Belgian relief ship

Notes to pages 5–7

Imo collided with the French munitions ship *Mont Blanc*. The *Mont Blanc* drifted ashore near Pier 6, just below the Richmond district of Halifax's North End. The ship exploded, killing more than 1,900 persons and wounding more than 4,000. More than 1,600 homes were destroyed and another 12,000 damaged. For scholarly studies of the explosion and its aftermath, see Ruffman and Howell, *Ground Zero* and Morton, *Ideal Surroundings*.

13 For a sketch of Murray's early career, see Brouwer, "Beyond 'Women's Work for Women,'" 65–95.

14 DUAPO, A-573, D. Fraser Harris to Dr A.S. MacKenzie, 6 November 1919.

15 To date, there have been a couple of brief treatments of the pathology laboratory. One, a thirty-year-old master's thesis, focuses on education and makes no mention of any of the early staff. The other is a brief article published more than thirty years ago by D.J. MacKenzie, the longtime laboratory director for whom the building housing the laboratory service at the Victoria General Hospital site of the Queen Elizabeth II Health Sciences Centre is named. MacKenzie, not surprisingly, focuses on the achievements and contributions of a select group of physicians and scientists and hardly mentions other staff. He does note that Low provided "faithful and efficient service" until her retirement in 1947. See Verma, "Medical Laboratory Technology Instruction in Nova Scotia"; and MacKenzie, "Origin and Development," 182.

16 PHAR, October 1926–September 1927, and 1928–29.

17 Prior to working for Harris, Low did "a great deal of technical work for Professor [A.G.] Nicholls." Harris complained that he did not have time to prepare the slides for this Histology class, and that his "time could be better occupied than with so mechanical a work." See DUAPO, A-605, Harris to MacKenzie, 27 July 1917.

18 DUAPO, A-573, A.S. MacKenzie to D. Fraser Harris, 27 July 1918.

19 See ibid., Harris to MacKenzie, 11 June 1919, 11 July 1919, 30 July 1919, and 6 November 1919.

20 Gagan and Gagan, *For Patients of Moderate Means*, 32, 34, 64–5.

21 All data come from the *New Brunswick Public Accounts*. The classifications were many, ranging from "lab assistant" (occasionally numbered ordinally) to "services." Entries under services include relief clerical work, laboratory work, and janitorial duties. It is not always possible to ascertain exactly what services were rendered.

22 See the final report of the Commission of Inquiry on the Blood System in Canada, and for a popular account, Picard, *The Gift of Death*.

23 One account from the point of view of a laboratory technologist is Marilyn Farrell, "The Walkerton Tragedy," 64–5.

24 There were other outbreaks of cryptosporidium in the 1990s. Two thousand people became sick in Cranbrook, British Columbia, and perhaps ten

to fifteen thousand in Kelowna, BC, during the summer of 1996. The largest outbreak, however, was in Milwaukee, Wisconsin, when one hundred people died and four hundred thousand became ill in 1993.
25 A number of books addressing these subjects have enjoyed considerable popularity. See for example Preston, *The Hot Zone*; Garrett, *The Coming Plague*. In the United States, Congress introduced the Clinical Laboratory Improvement Amendments in 1988 to improve the quality of laboratory work after public concern about AIDS brought the laboratory into focus. For a discussion, see Scarselletta, "The Infamous 'Lab Error.'"
26 This assessment on the spatial concept is based on oral interviews I have conducted with health care workers for several projects and from conversations with contemporary health care workers. For the architecture of hospitals, including some reference to laboratories, see Brandt and Sloane, "Of Beds and Benches," 290.
27 Barley and Orr, "The Neglected Workforce."
28 Keefe and Potosky, "Technical Dissonance," 54–5.
29 Whalley and Barley, "Technical Work," 23–4.
30 Warner, "Science in Medicine," 37. In part, Warner attributed this shift to the rise of the so-called "new social history" of medicine, which marked a significant shift in the content of medicine's historiography. Warner perceptively noted, however, that this same historiographic impulse "broadened the range of questions" asked about "science's place in medicine, such as its function as an ideology, a source of cultural authority, an agent of professional legitimation, and a tool for attaining social, economic, and political objectives." See also Warner, "Ideals of Science," 454–78.
31 Cunningham and Williams, "Introduction," 2.
32 Warner, "Science in Medicine," 39.
33 This trend predated the bacteriological and therapeutic advances of the 1870s and 1880s, a fact often omitted in discussions centred on the North American context. Arleen Tuchman suggested that in the middle of the nineteenth century, physicians in Baden "could hardly claim greater effectiveness than their competitors, but by arguing for stricter educational requirements based on increased work in the laboratory and clinic, they were promising to acquire those skills that would let them join the slowly emerging elite of scientists." Tuchman, *Science*, 3. For a North American perspective, see Warner, "The Rise and Fall of professional mystery," 133.
34 It is important to note that the laboratory as a site of authority has been strenuously critiqued. Hereditarian and environmentalist perspectives endured even after the discoveries of Pasteur and Koch. See Weindling, "Scientific Elites," 142–69. John Harley Warner has added that "the promise that laboratory science would transform medical practice rested, for a time, largely on optimistic faith, but that faith was not entirely blind." See Warner, "The rise and fall of professional mystery," 139.

35 Two examples will suffice. In a study of Miramichi Hospital, nearly every division or department within the hospital receives mention except for the laboratory facilities; see Gill, *75 Years of Caring*. This invisibility extends to the periodic anniversary celebrations found in newspaper supplements. The obligatory centennial retrospective of the Saint John General Hospital makes only fleeting reference to the laboratory, noting that "1954 saw the Provincial Laboratory moved from the hospital to a new building." See Supplement to the Saint John *Telegraph Journal*, 25 September 1967.
36 Bonner, *Medicine in Chicago 1850–1950*, 161. Charles E. Rosenberg succinctly addressed this point: "From a late-twentieth-century perspective, the resources of hospitals in the era of World War I may seem primitive, but they were impressive to contemporaries. Antiseptic surgery, the x-ray, the clinical laboratories represented a newly scientific and efficacious medicine – a medicine necessarily based in the hospital" ("Community and Communities," 9).
37 *Annual Report of the Hotel Dieu Hospital*, Chatham, New Brunswick, 1 August 1927 to 31 July 1928.
38 NBARMH, 31 October 1936.
39 Geison, *Private Science*, 49.
40 Bruce, *The Launching of Modern American Science*, 9, 13.
41 Bonner, *Medicine in Chicago*, 42. The emphasis is mine.
42 Cited in Rogers, *Dirt and Disease*, 22.
43 Shapin, *A Social History of Truth*, 378–9.
44 There are of course, exceptions. See, for example, Shapin, "The Invisible Technician," 554–63, and chapter 8 of his *Social History of Truth*; Mukerji, *A Fragile Power*, chapter 7; and Shaffer, "Astronomers Mark Time," 115–45. Theoretical issues of invisibility are addressed in Star, "Sociology of the Invisible," and Daniels, "Invisible Work." Technicians are of course mentioned in a large number of works, but largely in passing.
45 Shapin, *A Social History of Truth*, 126.
46 Hughes, *Men and Their Work*, and *The Sociological Eye*.
47 Freidson, *Professional Dominance*.
48 Etzioni, *The Semi-Professions*.
49 Witz, *Professions and Patriarchy*, 2–3. Elianne Riska and Katarina Wegar have also recently challenged gender-neutral accounts of the professional division of labour in health and medical care. See their edited collection *Gender, Work and Medicine*.
50 For a review of this literature, see Witz, *Professions and Patriarchy*, part one, 1–69. The relations between gender, patriarchy, capitalism, and technology have been explored in many works including Hartmann, "Capitalism, Patriarchy and Job Segregation by Sex"; Lown, "Not So Much a Factory, More a Form of Patriarchy"; Walby, *Patriarchy at Work*, and *Theorizing Patriarchy*; Cockburn, *Brothers: Male Dominance and Technological Change*, and *Machinery of Dominance*.

51 Witz, *Professions and Patriarchy*, 197.
52 Ibid., 168.
53 Ibid., ch. 6.
54 Ibid., 110–21, 173.
55 For comparisons, see Larkin, *Occupational Monopoly and Modern Medicine*, and Witz, *Professions and Patriarchy*, 169–72.
56 Larkin, *Occupational Monopoly*.
57 Baker, *Technology and Women's Work*; Cowan, *More Work for Mother*, and *A Social History of American Technology*.
58 Larson, *The Rise of Professionalism*.
59 Mitchinson, *Giving Birth in Canada*, 11.
60 Rosner, *A Once Charitable Enterprise*; Vogel, *The Invention of the Modern Hospital*; Latour and Woolgar, *Laboratory Life*; Howell, *Technology in the Hospital*.
61 McKay, "A Note on 'Region,'" 92.
62 Gagan and Gagan, *For Patients of Moderate Means*, 46, 89; Godfrey, "Private and Government Funding," 22–3.
63 A selection of historical studies on the medical sciences would include Coleman and Holmes, eds., *The Investigative Enterprise*; Collard, *The Development of Microbiology*; Cunningham and Williams, eds., *The Laboratory Revolution in Medicine*; Fye, *The Development of American Physiology*; Geison, ed., *Physiology in the American Context*; Holmes, *Claude Bernard and Animal Chemistry*; Jacyna "The Laboratory and the Clinic"; Kohler, *From Medical Chemistry to Biochemistry*; Mazumdar, *Immunology 1930–1980*.
64 For a consideration of where scientific and technical workers fit within the class structure, see Gorz, "Technology, Technicians and Class Struggle," and Mandel, *Late Capitalism*.
65 There are few studies that take a national perspective, for the reasons explicated earlier, and we still lack a synthesis of the history of Canadian health care. Some terrific recent studies are beginning to fill major gaps in the historiography, including fresh analyses by David Gagan and Rosemary Gagan on hospitals (*For Patients of Moderate Means*), Wendy Mitchinson on childbirth (*Giving Birth in Canada*), and Kathryn McPherson on nursing (*Bedside Matters*). There are also many exemplary studies that are more geographically specific that have enriched our understanding of health care's past. See, for example Moran, *Committed to the State Asylum*; Reaume, *Remembrance of Patients Past*; Adams, *A Dentist and a Gentleman*.
66 This is most clearly illustrated through commemorative hospital histories, which are often rich in empirical detail but limited in scope and analytical content. For an example of this genre, see MacLellan, *A History of the Moncton Hospital*. For an interesting discussion of writing commissioned histories of hospitals, see Heaman, "Review of Christopher J. Rutty," 163–8. Connor, "Hospital History in Canada and the United States," offers a good review of the historiography to 1990, but it is in need of updating.

67 Bator, "Saving Lives," 12.
68 The broadest Canadian views are Bator, "Saving Lives"; MacDougall, *Activists and Advocates*; Defries, *The Federal and Provincial Health Services in Canada*. For public health in particular settings, see Artibise, *Winnipeg*, especially chapter 13; Copp, *The Anatomy of Poverty*; Andrews, "The Best Advertisement a City Can Have," 19–27.
69 Cassel, "Public Health in Canada," 287–8; McCuaig, *The Weariness, the Fever, and the Fret*, 6.
70 The federal government's role in providing medical services to First Nation communities has been explored in several recent studies, including Kelm, *Colonizing Bodies*, and Lux, *Medicine That Walks*.

CHAPTER TWO

1 For Canada, see Bator, "Saving Lives"; MacDougall, *Activists and Advocates*; Artibise, *Winnipeg: A Social History*, especially chapter 13; and Defries, *The Development of Public Health in Canada*.
2 McGhie, "The Laboratory in Relation to Public Health," 38.
3 Gagnon, "Notes," 220.
4 Reiser, *Medicine and the Reign of Technology*, 141–2. Locally, some physicians in Halifax remained unconvinced during the 1890s of the utility of laboratory testing for diphtheria. Nevertheless, by the end of the century, there was widespread agreement about the need "for a certain diagnosis." By 1899 the laboratory reported making more tests in a single year than it had in all previous years combined. JHA 1899, Appendix 14.
5 JHA 1897, Appendix 14.
6 Brandt, *No Magic Bullet*, 40; and Cassels, *The Secret Plague*, 19.
7 MacKenzie, "The Origin and Development of a Medical Laboratory Service in Halifax," 179–84. In "The Development of Public Health in Nova Scotia," Campbell and Scammell suggest that "a Provincial Diagnostic Laboratory" was established in 1894.
8 "Report of the Provincial Board of Health" in JHA 1896, Appendix 14. The terms of reference for the committee, which comprised Drs. Farrell, Sinclair, and Reid, may be found in JHA 1897, Appendix 14. The concern over food and water purity was common in the public health movement everywhere. See, for example, MacDougall, "Health is Wealth," 313–14.
9 JHA 1897, Appendix 14. One early history suggested that "equipment was of the most elementary character and for a long time the only microscope was the property of the director. It is still recalled how the lack of an incubator made necessary the employment of human incubators and house surgeons at the hospital slept with their culture tubes in the pockets of their night attire." Campbell and Scammell, "Development of Public Health in Nova Scotia," 233.

10 JHA 1902, Appendix 16.
11 *MMN* 15 (1903): 108.
12 *MMN* 12 (1900): 376, and NSARM RG25, Series C, Vol. 1, Andrew Halliday to A.P. Reid, 29 January 1901. Details of Reid's career may be found in Howell, "Medical Science and Social Criticism," and Linkletter, "An Open Door."
13 NSARM RG25, Series C, Vol. 1, A.P. Reid to Hon. George H. Murray, 11 April 1901.
14 *MMN*, 13 (1901): 297, 327.
15 JHA 1903, Appendix 16; JHA 1904, Appendix 16, and *MMN* 15 (1903): 100–1, 108–9. This issue contains an editorial on Halliday and his contribution to the medical profession and public health as well as an obituary.
16 *MMN* 15 (1903): 102.
17 JHA 1903, Appendix 16, and *MMN* 15 (May 1903): 181.
18 JHA 1910, Appendix 16.
19 JHA 1911, Appendix 16 and JHA 1912, Appendix 16.
20 NSARM RG25 Series B, Vol. 3, BOC, 14 January and 18 February 1911.
21 Ibid., BOC, 7 July 1911. The committee reported that the proposed laboratory would cost about $9,000. The minutes note, however, that "after some discussion the matter was referred back to [the architect] for plans of a wood building with estimates of cost of same." While committed to improving the laboratory service, it is clear that a measure of economy was to be maintained.
22 VGHL, W.W. Kenney to Dr John A. Hornsby, 28 October 1914, and W.W. Kenney to Dr Winder H. Smith, 21 September 1914. See also MacKenzie, "Laboratory Service in Halifax," 181–2. The laboratories have stayed close to their Morris Street location until the present day. Most laboratory services connected with the Victoria General Hospital or the Province were located in the Pathological Institute, which was modified in the 1920s and the 1960s. With the construction of the D.J. MacKenzie Building, many of the Victoria General's labs were moved.
23 VGHL, W.W. Kenney to Dr John A. Hornsby, 28 October 1914. See also ibid., Kenney to Dr M.C. Archibald, 29 May 1915.
24 VGHL, W.W. Kenney to M.A. Lindsay, 3 May 1911.
25 NSARM, RG25, Series B, Vol. 3, BOC, 25 October 1910; 3 December 1910; 1 April 1911; 29 April 1911, and 13 May 1911.
26 VGHL, W.W. Kenney to M.A. Lindsay, 17 May 1911.
27 VGHL, W.W. Kenney to Reuben O'Brien, 15 June 1914, and PHAR, October 1913–September 1914. The *Empress of Ireland* collided with a Norwegian freighter and sank in the St Lawrence River, killing 1,012 passengers.
28 VGHL, W.W. Kenney to Dr Winder H. Smith, 21 September 1914.
29 NSARM, RG25, Series B, Vol. 3, BOC, 8 June 1914.

30 The actual process of finding a candidate was multifaceted. Kenney corresponded with no less a figure than Dr J.G. Adami, the esteemed Montreal pathologist. President MacKenzie of Dalhousie also put forward several names, as did contacts in Boston, New York, and Toronto. At the same time, several letters were received from British physicians interested in the pathologist position. Finally, Dr W.H. Hattie was sent to Ontario to find a pathologist. While there, he identified Nicholls as the best candidate. This account was reconstructed using BOC entries from the late spring, summer, and fall of 1914.
31 *NSMB* 25, no. 4 (April 1946).
32 PHAR, October 1914–September 1915.
33 NSARM, RG25, Series B, Vol. 3, BOC, 16 October 1914. Salary was an issue. Nicholls wanted $3,000 but eventually settled for $2,500 and a $300 moving allowance. See ibid., 26 October 1914, 11 November 1914, and 21 November 1914; VGHL, W.W. Kenney to A.E. [*sic*] Nicholls, 31 October 1914; DUAPO, File 359, A.S. MacKenzie to A.G. Nicholls, 13 January 1915. Dalhousie was unwilling to contemplate anything more than its $500 contribution, so responsibility for the increase fell to the provincial government.
34 McKay, "The Stillborn Triumph of Progressive Reform," 293.
35 Beck, *Politics of Nova Scotia, Vol. 2*, 94.
36 Penney, "Marked for Slaughter," 27–51; Murray, "The Visit of Abraham Flexner," 34–41. With the support of the Carnegie Foundation, Abraham Flexner investigated the state of medical education in the United States and Canada. Flexner's report severely criticized the Canadian medical schools at Queen's, Laval, Western Ontario, and in Halifax. The influential report did much to document the deficiencies in medical education, but the reform of medical education was already underway as Penney concludes. By the time Flexner completed his report in 1910, the number of medical schools in the United States had already decreased by twenty-five percent.
37 The antituberculosis association was created on 18 November 1909 and encompassed Pictou, Antigonish, and Guysborough counties. See Cox and MacLeod, *Consumption*; Penney, *Tuberculosis in Nova Scotia*; McCuaig, "'From Social Reform to Social Service,'" 480–501.
38 The mission was founded in 1868 by Edward Jost and provided important services to women, including educational programs and child cure. See Simmons, "'Helping the Poorer Sisters,'" 3–27.
39 Here the work of E.R. Forbes is most important. See Forbes, "Prohibition and the Social Gospel," 11–36, and "The Ideas of Carol Bacchi," 119–26.
40 For PEI, see Baldwin, "Volunteers in Action," 121–47. New Brunswick is discussed below.
41 Twohig, "Public Health in Industrial Cape Breton, 1900–1930s."
42 Twohig, *Challenge and Change*, 9–11.

43 The literature on the deindustrialization of the Maritimes is vast. A good summary of the literature is chapter 6 of Reid, *Six Crucial Decades*.

44 There are many examples, including the creation of the public health nursing program at Dalhousie University, which depended upon support from the Red Cross, the role of the Canadian Tuberculosis Association in financing the public health department on Prince Edward Island, and the extensive activities of the Rockefeller Foundation. For details, see Twohig, "'To produce an article,'" 26–41, and *Challenge and Change*, ch. 1; Baldwin, "Volunteers in Action"; and Reid, "Health, Education, Economy."

45 MMN, entries for 2 August, 4 October, and 19 October 1909. *Chronicle* [Halifax], 28 November 1910.

46 NSARM, RG25, Series B, Vol. 3, BOC, entries for 24 June and 18 July 1911.

47 MMN, 20 June 1911.

48 NSARM, RG25, Series B, Vol. 3, BOC, entry for 10 June 1911.

49 Ibid., entry for 5 August 1911. This figure remained unchanged while the laboratory stayed under the auspices of the VG. See VGHL, Kenney to MacKenzie for the following dates: 25 May 1915, 7 July 1916, 11 May 1918, 23 May 1919, 26 May 1920, 5 July 1921, and 16 May 1924.

50 NSARM, RG25, Series B, Vol. 3, BOC, entries for 5 July, 2 August, and 20 December 1913. Occasionally, the laboratory was billed for equipment that belonged to Dalhousie. Kenney wrote to one supplier that "Dalhousie University and Victoria General Hospital are two distinct institutions and have nothing in common whatsoever, at least so far as revenue and expenditure is concerned" (VGHL, Kenney to J.F. Hartz Co., Ltd, Toronto, 26 February 1917).

51 See VGHL, Kenney to MacKenzie, 5 August 1913. This clause was strictly interpreted. When Dr D. Fraser Harris of the Department of Physiology wanted to establish a lecture room in the facility, he was rebuked by Kenney. "Teaching" clearly meant work at the bench and did not include didactic lectures. See VGHL, Kenney to Harris, 14 March 1916. President MacKenzie issued a notice at the hospital's request prohibiting Dalhousie students from the laboratory "except when classes are being held during the regular scheduled lecture or laboratory hours." DUAPO, A-817, MacKenzie to Kenney, 12 September 1914.

52 NSARM, RG25, Series B, Vol. 3, BOC, 30 June 1916.

53 VGHL, Kenney to Chisholm, 28 March 1925.

54 NSARM, RG25, Series B, Vol. 3, BOC, 27 March 1925.

55 Ibid., 23 September 1925, 7 May, 16 July, 16 August, 26 August, and 28 September 1926. In July the board suggested two models: (1) that the public health department assume responsibility for all laboratory work and staff and charging the hospital for any services rendered, or (2) that the hospital assume responsibility for the work and staff and bill the province for

the public health work. When Drs Jost and Chisholm were presented with the second option, they objected strongly and left a meeting with the board.
56 DUAPO, A-817, A.C. Jost to A.S. MacKenzie, 20 September and 21 September 1926. NSARM, RG25, Series B, Vol. 3, BOC, 28 September and 16 October 1926.
57 Ibid.
58 DUAPO, A-572, [Gilbert Stairs?] to Dr A.W.H. Lindsay, 4 July 1911. This letter concerns the establishment of a provisional faculty of medicine within Dalhousie University. It notes, "With regard to Pathology and Bacteriology, the Governors have – in accordance with the previous recommendation of the Faculty – appointed as Professor and head of this department Dr M.A. Lindsay, recently appointed by the Hospital Commission as Pathologist to the Victoria General Hospital." See also minutes of the committee appointed by the governors of Dalhousie University to confer with the representatives of the medical profession with regard to the organization of a teaching medical faculty, 18 May 1911, in the same file. These minutes record "That the person appointed by the Hospital Commission as Pathologist be selected as Professor of Pathology and Bacteriology."
59 DUAPO, A-995, A.S. MacKenzie to Dr R.F. Rutten, Department of Chemistry, McGill University, 28 May 1919.
60 NSARM, RG25, Series B, Vol. 3, BOC, 18 June, 3 August, 21 December 1928, and 26 January and 11 September 1929.
61 DUAPO, A-995, A.S. MacKenzie to Dr R.M. Pearce, 19 June 1928.
62 NSMB 5, no. 12 (December 1926): 31, and PHAR, October 1926–September 1927.
63 A.G. Nicholls, "Report on the Work of the Public Health Laboratory for the Year Ending September 30, 1921," in PHAR, October 1920–September 1921, and VGHL, W.W. Kenney to Dr A. Stanley MacKenzie, 7 September 1921. In his letter, Kenney wrote that the board of commissioners "are quite agreeable to your [Dalhousie's] proposal that Dr D.J. MacKenzie be permitted to work under and with Dr Nichols [sic] in doing pathological work in the hospital laboratory. The Board assumes however that in doing this, any expense incurred beyond what would be necessary in carrying on the hospital work, would not be a charge against the institution."
64 A.G. Nicholls, "Report on the Work of the Public Health Laboratory for the Year Ending September 30, 1922," in PHAR, October 1921–September 1922, and "Report ... for the Year Ending September 30, 1923," in ibid., October 1922–September 1923; and VGHL, W.W. Kenney to Dr A.G. Nicholls, 17 June 1922.
65 Interview with Edna Williams, 23 April 1996, and interview with Rose Phillips, 22 April 1996.
66 DUAPO, MS-2-3, A-575, R.J. Bean, Secretary of the Medical Faculty, to Carleton M. Stanley, 2 March 1932.

67 PHAR, 1925–26, 22, and PHAR, 1926–27, 22–3. It is important to note that this increase was in specimens and not tests and is therefore unrelated to the dual testing of blood samples with both the Wassermann and the Kahn tests. The PHAR for 1927–28 (p. 21) estimated that the dual testing did account for over two thousand tests.

68 See PHAR, October 1930–September 1931, 7. For the five-year interval 1921–26, public health work increased by 11.3 percent. D.J. MacKenzie suggested that the increase was principally due to venereal disease work and sputum examinations, although he acknowledged that this growth would not continue. He did suggest, however, that the province should be conducting twice as many sputum examinations for tuberculosis in the fight against that disease. PHAR, October 1931–September 1932, 22.

69 Ibid., 1928–29, 23.

70 PANB RS 136, Records of the Deputy Minister of Health, File A.1a, "To the Premier and Gentlemen of the Executive Council," n.d.

71 Hayter, "New Brunswick's Remarkable Dr Roberts," 1,637–9.

72 PANB RS 136, Records of the Deputy Minister of Health, File E5a, William F. Roberts to Rev. Sister Superior Walsh, 1 October 1919.

73 PANB RS 136, Records of the Deputy Minister of Health, File A.1a, George G. Melvin, "Five Years of the NB Public Health Act," 20 November 1923, and NBARMH, 31 October 1918.

74 *New York Times*, 18 April 1934; and Rogers, *Dirt and Disease*.

75 NBARMH, 31 October 1918. In exchange for this space, the General Hospital received $500 from the Province. As well, the provincial laboratory "was to provide a general pathologic, bacteriologic, and serologic service to the hospital, and the Chief of Laboratories was to be considered the pathologist to the General Public Hospital."

76 Ibid.

77 Ibid., 31 October 1921.

78 Ibid.

79 PANB RS 136, Records of the Deputy Minister of Health, Files E5b, E30b, E8, and E7b. These files all contain questionnaires that were administered to hospitals annually and that asked, among other things, about laboratory facilities.

80 Ibid., File E5a, Sr Walsh to W.F. Roberts, 13 August 1919.

81 Ibid., File E5a, Roberts to Walsh, 1 October 1919.

82 Ibid.; and File E24a, Dr H.L. Abramson to Dr L.G. Pinault, 25 September 1922.

83 NBARMH, 31 October 1919.

84 Ibid., 31 October 1918.

85 PANB RS 136, Records of the Deputy Minister of Health, File Q1, H.L. Abramson to "Dear Doctor," 5 November 1923.

86 NSARM RG25 Series B, Vol. 2(B), MMB, 11 October 1909.

87 Most of the laboratory equipment and supplies were ordered from firms in Toronto, the United States, or Europe. See C.E. Puttner to Bausch and Lomb Optical Co., 31 July 1913 and 28 May 1918; W.W. Kenney to J.B. Prescott and Son, Webster, Massachusetts, 6 March 1914; Kenney to Lederle Antitoxin Laboratories, 15 August 1919; Kenney to Dr A.G. Nicholls, 20 April 1923, all in VGHL. Local businesses were sometimes used. In 1915, for example, Charles Coburn of Hantsport supplied the laboratory with guinea pigs. VGHL, W.W. Kenney to Charles D. Coburn, Hantsport, 13 October 1915.
88 MMN 8, no. 11 (November 1896): 359.
89 JHA 1899, Appendix 14.
90 JHA 1901, Appendix 12; VGHL, L.M. Murray to S.N. Miller, 16 November 1914, Murray to Miller, 30 November 1914, and L.M. Murray to A.K. Roy, 3 November 1914.
91 VGHL, W.W. Kenney to Dr F.C.L., New Ross, 28 October 1913.
92 JHA 1905, Appendix 16.
93 W.E. MacLellan, Post Office Inspector to L.M. Murray, December 1, 1905, reprinted in JHA 1905, Appendix 16.
94 JHA 1910, Appendix 16, and PHAR, October 1931–September 1932, 22–3.
95 JHA 1907, Appendix 16. Murray reported that for most specimens forwarded by private individuals, the name of a physician was supplied and the results reported.
96 This anecdote is recounted in Overduin, *People and Ideas*, 40–1.
97 Reid, "Public Health," in MMN 13 (August 1901): 285–6. Reid originally made the comments in an address before a meeting of the Maritime Medical Association, 4 July 1901.
98 JHA 1908, Appendix 13.
99 JHA 1900, Appendix 12.
100 On vital statistics, see Emery, *Facts of Life*, and, for Nova Scotia specifically, Dunlop, "The Collection of Vital Statistics in Nova Scotia." The most outstanding Canadian analysis of the social and political use of such statistical data is Curtis, *The Politics of Population*.
101 JHA 1898, Appendix 14.
102 JHA 1905, Appendix 16.
103 MMN, 8, no. 11 (November 1896): 358–9. Widal published his agglutination technique in 1896 and that same year, Hattie performed over one hundred of them in Halifax. This suggests that the lag-time in adopting and implementing new methods was very short in some instances. See MacKenzie, "The Origin and Development of a Medical Laboratory Service in Halifax," 180. Physicians, it should be noted, were often reluctant participants in the exchange of information. In 1899 William Hattie lamented:

> I have again to regret that it so often happens that once we have furnished a report on a case, we hear nothing more of it. It is such a small matter to drop a post-card

stating whether the ultimate history of a case agreed with the report that the laboratory had furnished, that physicians are very apt to overlook it altogether. Nevertheless unless this is done it is impossible to form any true idea as to the accuracy of the work ... I trust that in future those who receive assistance from the laboratory will not neglect to inform us to the accuracy of our reports. (JHA 1899, Appendix 14)

104 Nicholson, *Laboratory Medicine*, 227.
105 Very little of this correspondence has been preserved and there is indirect evidence that the results were communicated by telephone (see VGHL, W.W. Kenney to B.S. Bishop, 18 August 1911). However, a very good run of letters may be found in the VGHL for November and December 1914, which reveals the range of samples being examined and confirms that samples were being received from throughout the province.
106 VGHL, L.M. Murray to L. Rockwell, 31 October 1914, and Murray to Dr W. Rockwell, 3 November 1914.
107 NSMB, 10, no. 10 (October 1931). This statement accompanied the provincial pathologist's monthly reports in the *Bulletin*.
108 PHAR, October 1915–September 1916.
109 An article from 1902 suggested that "the general practitioner is apt to look upon microscopical work as of scientific rather than practical value." Fraser, "The Microscope in Diagnosis," 162. The Presidential Address of Dr L.R. Morse in NSMB 1 (December 1922): 6 reaffirmed the paramount importance of clinical judgment. For a contrary opinion, see Mack, "The General Practitioner," 6–10.
110 Warner, "Science in Medicine," 45.
111 Ibid.
112 Morantz-Sanchez, *Sympathy and Science*, 236.

CHAPTER THREE

1 PHAR, October 1919–September 1920, 18.
2 Interview with Ellen Robinson, conducted by Peter L. Twohig, 29 April 1996.
3 Interview with Edna Williams, conducted by Peter L. Twohig, 23 April 1996.
4 Ibid.
5 JHA 1898, Appendix 14, and JHA 1899, Appendix 14.
6 The specimens included 65 sputum specimens for TB testing, 43 throat swabs for diphtheria testing, 138 blood samples for typhoid, and twenty other samples for various tests. The specimens were forwarded by 120 physicians throughout the province. JHA 1898, Appendix 14.
7 PHAR, October 1919–September 1920, 16.

8 PHAR, October 1920–September 1921, 23. As always, much of the testing resulted from public health efforts in the province. However, in his report, A.G. Nicholls wrote, "Part of the increase in the amount of material submitted for examination is due to the opening up of new hospitals and clinics in the Province and the activities of the Massachusetts Halifax Health Commission in Halifax." Comments on the expansion of public health work are found throughout the PHAR; see for example October 1923–September 1924 or October 1924–September 1925. For the entire period, 1921–26, public health work increased by 11.3 percent.

9 The examples that follow come from Nicholson, *Laboratory Medicine*, published in 1930. Nicholson was a professor of pathology at the University of Manitoba and a pathology assistant at the Winnipeg General Hospital when the book was published. This edition was selected because it was more or less in the middle of the period under study and enjoyed a good reputation.

10 Formalin was added to a typhoid culture to reduce it to twenty-five percent strength of the total culture. This killed the organisms but the bacilli retained their agglutination properties. The solution could be stored for several months in an ice-chest.

11 All of these data are drawn from Nova Scotia's Department of Health Annual Report. The reports occasionally gave the number of tests or the number of samples. Both indicate the expansion of the work, however.

12 Collard, *The Development of Microbiology*.

13 PHAR, October 1913–September 1914, 10. Diphtheria antitoxin enjoys a prominent place in Canadian history, leading ultimately to the creation of the Connaught Laboratories in Toronto. John Gerald Fitzgerald, an associate professor of hygiene at the University of Toronto, began producing antitoxin in 1913. Bator has suggested that "the large number of children dying from diphtheria in Canada, despite the existence of a method of treating the disease with antitoxin, angered and baffled many doctors" (*Within Reach of Everyone*, 1). But the value of providing antitoxin to physicians or health authorities is, of course, dubious. Barbara Gutmann Rosenkrantz has argued that it is not clear that "the splendid practice of making diphtheria antitoxin available without charge to the physician who treated the poor, as in New York City, or entirely free, as in Massachusetts, had any significant effect on the immunization of patients" ("Cart before Horse," 71). See also Rutty, "Poliomyelitis in Canada," 52–3.

14 Nicholson, *Laboratory Medicine*, and Todd and Sanford, *Clinical Diagnosis by Laboratory Methods*, 820.

15 PHAR, October 1915–September 1916, 15.

16 MMB, NSARM RG25 Series B, Vol. 2 (B), MMB, 12 January 1916.

17 Rogers, *Dirt and Disease*, 4.

18 VGHL, W.W. Kenney to W.H. Hattie, 10 January 1916; Kenney to Mrs Margaret Cameron, 18 January 1916; Kenney to Dr Seymour MacKenzie, 21 January 1916; and Kenney to Dr W.B. Almon, 24 January 1916.
19 JHA 1902, Appendix 16.
20 JHA 1903, Appendix 16.
21 JHA 1908, Appendix 13. Noting this development, Dr L.M. Murray, the director, added that the "subject of animal diseases in their relationship to Public Health, will, probably, in a short time, become of such importance as to demand the special attention of [the Department of Health]" (ibid.).
22 See MacDougall, "Health is Wealth," 297; McCuaig, *The Weariness*, 159.
23 PHAR, October 1915–September 1916, 12. While acknowledging the importance of pure milk, the report goes on to suggest that "even more potent factors are the wretched housing conditions ... and the poverty of a very considerable proportion of the people" (ibid.)
24 MMN, 18, no. 12 (December 1906): 446. Incidentally, readers may be interested to note that the same issue of the MMN contains a paper read before the Saint John Medical Society on the milk question. Although it makes very little mention of the local context, it does reprint the milk regulations for that city. See Daniel, "The Milk Supply and Its Control."
25 Bator, "Saving Lives," 156.
26 Hayter, "New Brunswick's Remarkable Dr Roberts," 1,638; McCuaig, *The Weariness*, 169.
27 JHA 1905, Appendix 16.
28 Deutsch, "Menace in the Medical Labs," 89.
29 Cassels, *The Secret Plague*, 147–50.
30 Portions of Bryce's career are covered in Sproule-Jones, "Crusading for the Forgotten," 199–224.
31 Ibid.
32 For Ontario, see Bator, "Saving Lives," 255–6.
33 Cassels, *The Secret Plague*.
34 Ibid., 163–9.
35 Cassels, passim, and Buckley and McGinnis, "Venereal Disease," 353.
36 Cassels, *The Secret Plague*, 169.
37 PHAR, October 1919–September 1920, 10.
38 Ibid., 9–10. In Halifax, Admiralty House was opened for the treatment of venereal diseases. See VGHL, W.W. Kenney to Dr J.A. Doull, Provincial Inspector of Health, 17 March 1920.
39 Buckley and McGinnis, "Venereal Disease," 349–50.
40 Judith R. Walkowitz noted that interest in Britain's Contagious Diseases Acts also focused narrowly on the "cultural importance" of the legislation and failed to consider "the administrative machinery and the medical

technology that facilitated their operation." Walkowitz, *Prostitution and Victorian Society*, 69.
41 Brandt, *No Magic Bullet*, chapter 4.
42 PHAR, 1919–20 to 1929–30. The laboratory reported doing some 159 Kahn tests in 1925–26, and this number grew to over 2,700 the next year, outpacing the 2,369 Wassermann's tests that had been completed. Thereafter, only Kahn's were conducted, topping 4,000 in 1928–29, almost 7,000 in 1930–31, and 9,000 in 1933–34, reaching 10,000 in 1934–35.
43 For a brief description of the Wassermann reaction, see *Maritime Medical News*, 22, no. 1 (January 1910): 2. See also MacKenzie, "The Origin and Development of a Medical Laboratory Service," 182.
44 Nicholson, *Laboratory Medicine*, 198.
45 Ibid.
46 Ibid., 378, 385. Not surprisingly, hand cleaning was also a significant theme in the laboratory manuals. Fingernails were particularly suspect. They were to be kept "short, neat and clean," maintained through a regimen of soap, water, and a nail brush. The resulting good lather would destroy any diphtheria bacilli, streptococci, pneumococci, meningococci, or gonococci. Workers were cautioned against handling chemical or infective material as the skin could never be entirely sterilized. Soaps with antiseptics were considered to be of "questionable value" because they might give workers a "false sense of security." Nicholson concluded that "it is as important for the laboratory worker as for the practising doctor to have socially clean hands at all times."
47 PHAR, December 1937–November 1938, 34.
48 Nicholson, *Laboratory Medicine*, 198.
49 Ibid.
50 JHA 1909, Appendix 16.
51 For the Canadian context on polio, see Rutty, "The Middle Class Plague," 277–314.
52 On this early effort, see Benison, "Poliomyelitis," 74–92.
53 Rogers, *Polio before FDR*, 57–9.
54 Ibid., 73–5.
55 MacKenzie himself published an article on the disease; see "Some Phases of Poliomyelitis," 415–17.
56 PHAR, October 1928–September 1929, 23.
57 For a discussion of convalescent serum, see Paul, *A History of Poliomyelitis*, 198.
58 PHAR, October 1929–September 1930, 9.
59 NSMB 10, no. 10 (October 1931): 720. Interestingly, this notice appeared under the "Personal Interest Note" section.
60 PHAR, October 1930–September 1931, 23.
61 PHAR, October 1931–September 1932, 13.

62 PHAR, October 1933–September 1934, 10, and October 1934–November 1935, 9. Mortality for polio during the middle years of the 1930s was small. In 1932 four people died, six in 1933, four in 1934, two in 1935, and one in 1936. It nevertheless engendered tremendous fear among the populace. See PHAR, December 1936–November 1937.

63 PHAR, December 1935–November 1936, 12, and December 1936–November 1937, 12. See also Paul, *A History of Poliomyelitis*, 248.

64 MacKenzie, "The Origin and Development of a Medical Laboratory Service," 182. The discovery of insulin is, of course, a well-known and significant story in Canadian medical history. The details may be found in Bliss, *The Discovery of Insulin*, and *Banting: A Biography*.

65 DUAPO Staff Files, File 467, George M. Murphy, Minister of Public Health, to Carleton Stanley, 22 September 1931; Howell, *A Century of Care*, 72; PHAR, October 1932–September 1933, 6, October 1934–November 1935, 13. One of the stated objectives of the Cancer Clinic was the study of tumours to establish their type and location. The *Report of the Department of Public Health* for 1932–33 noted that the laboratory "now has its doors wide open" for tissue work. PHAR, October 1932–September 1933, 6.

66 NSMB 10 (October 1931): 692. The *Bulletin* reproduced a letter from George H. Murphy, minister of Health, dated 26 September 1931. Following this letter, the provincial pathologist's monthly returns of work performed in the laboratory were reported in subsequent issues of the *Bulletin*. The reports detailed the number of malignant, simple, and suspicious tumours, "other conditions," as well as the number of tissue samples awaiting section.

67 NBARMH, 31 October 1918, 33. Interestingly, Abramson made this request after only a few months of work. The laboratory began operations officially on 1 June 1918.

68 NBARMH, 31 October 1921, 4–5.

69 *New Brunswick Public Accounts*. LeBrun earned $123.63 for her work in 1921, apparently receiving a daily wage. This continued the next year, until she was finally put on salary mid-way through fiscal year 1921–22. For the first twenty-five weeks she earned almost $491, while for the last five months of the year she earned $416.67, suggesting that the move to "salary" did little to increase her pay envelope.

70 NBARMH, Annual Report of the Bureau of Laboratories Year Ending 31 October 1925, 4.

71 Ibid., Year Ending 31 October 1918, 30.

72 Ibid., Year Ending 31 October 1925, 3. In the first year, more than one thousand Kahn tests were performed, several hundred purely for experimental reasons. In Nova Scotia, a comparison was made between the Kahn and Wassermann tests based on two thousand consecutive cases. The Kahn test was determined to be "more suitable" and the Wassermann test was

discontinued in August 1927. See PHAR, 1926–27, 22–3. The annual report noted clearly that the Wassermann tests ceased in August, although there may have been some questioning of this decision. The provincial health officer, Dr A.C. Jost, urged that the medical staff of the Victoria General be asked about the decision to discontinue Wassermann tests. The minutes of the Medical board record only that the board "could not speak definitely on the matter at the present time and was not justified in making any recommendation." MMB, 15 December 1927.

73 NBARMH, Annual Report of the Bureau of Laboratories Year Ending 31 October 1931, 17.
74 Ibid., Year Ending 31 October 1929, 5. The Saint John laboratory also performed other tests occasionally for PEI.
75 This description is based on Nicholson, *Laboratory Medicine*, 276. No statistics were kept on the number of urine tests completed each year at the Halifax laboratory. They were routine and of little interest to the public health authority that generated statistics on public health tests. The exception to this was when work was conducted for the Massachusetts-Halifax Health Commission. In 1924–25 the Commission asked the laboratory to complete 633 routine examinations and 501 examinations for sugar and albumin. In New Brunswick routine tests were similarly not reported on a regular basis in departmental annual reports.
76 NBARMH Annual Report of the Bureau of Laboratories Year Ending 31 October 1923, 3.
77 Ibid., Year Ending 31 October 1935, 3.
78 Ibid., Year Ending 31 October 1919, 35–6.
79 Ibid.
80 Ibid.
81 NBARMH, 31 October 1928, 15, and 31 October 1929, 5.
82 NBARMH, 31 October 1931, 10, and ibid., Annual Report of the Bureau of Laboratories Year Ending 31 October 1932, 6.
83 These numbers are based upon data published in the annual reports and reflect the number of tests, not the number of samples. The totals are my own. Despite the imperfections of the data, they do serve to illustrate the significant growth in the work of the laboratory.
84 VGHL, G.A. McIntosh to Raymond E. Devillez, 3 June 1924.
85 VGHL, G.A. McIntosh to Dr T.R. Johnson, 30 June 1924.
86 VGHL, W.W. Kenney to B. Franklin Royer, 12 January 1920.
87 VGHL, W.W. Kenney to Massachusetts-Halifax Health Commission, 26 June 1924. Another case saw a female patient undergo a Wassermann test, urine tests, and an x-ray of the gastrointestinal tract. A blood examination revealed slight anemia, and she was treated with alkalines for her stomach condition. Reporting to her physician, G.A. MacIntosh said that when discharged "she was recommended to be kept on alkalines with arsenic

and iron tonic and freedom from work." See VGHL, G.A. MacIntosh to J.R. MacLeod, 26 August 1925.
88 VGHL, G.A. McIntosh to Assistant Unit Medical Director, Camp Hill Hospital, 5 January 1925.
89 MacKenzie, "Origin and Development," 182.
90 This discussion is largely based on Reiser, *Medicine and the Reign of Technology*, 183–6.
91 Reiser, *Medicine*, 185. See also Scarselletta, "The Infamous 'Lab Error.'"
92 Reiser, *Medicine*, 185–6. One is struck by the similarity in Reiser's analysis with that of the famous Flexner report of 1910 on medical education. The old account saw eager young American physicians carry the latest concepts of medical education and science from Germany to Johns Hopkins and a select few other institutions. The same account notes the efforts of the American Medical Association in pressing for higher standards.
 The final blow in this account comes with the Flexner report, which initiated a period in which medical schools became fewer, smaller, and more scientific.
93 PHAR, October 1924–September 1925, 22.
94 NBARMH Annual Report of the Bureau of Laboratories Year Ending 31 October 1938, 4.
95 NSARM, RG25, Series B, Vol. 3, BOC, 16 April 1927.
96 D.J. MacKenzie, "Report of the Work of the Public Health Laboratory," in PHAR, October 1928–September 1929, 24.
97 Ibid.
98 PHAR, October 1929–September 1930, 15–6.
99 PHAR, December 1936–November 1937, 32–3.
100 PHAR, October 1929–September 1930, 7.
101 In a statement typical of annual reports in the later 1920s and 1930s, the annual report for 1934–35 declared that the laboratory was "absolutely essential to the proper functioning" of the health department. PHAR, October 1934–September 1935, 11.
102 PHAR, October 1933–September 1934, 12.
103 PHAR, October 1934–November 1935, 11.
104 PHAR, December 1935–November 1936, 15.

CHAPTER FOUR

1 Rosen, *The Structure of American Medical Practice*, 46.
2 Urquhart, ed., *Historical Statistics of Canada*, 49.
3 See Cameron, *"And Martha Served."*
4 St Martha's Hospital Fund Campaign, "The Story of St Martha's Hospital," 5.
5 MacDonald, "Golden Gleanings".

6 Pothier, *Mary Ann Watson*, 33-4.
7 Cogswell, *Western Kings Memorial Hospital*.
8 *Moncton Daily Times*, 11 June 1968 [Moncton Hospital Anniversary Supplement]; Gill, *75 Years of Caring: A History of the Miramichi Hospital*; and *75 Years of Caring: St Joseph's Hospital*.
9 For the history of some of these institutions, see Francis, "The Development of the Lunatic Asylum in the Maritime Provinces," 23-38.
10 McPherson, "Nurses and Nursing," 14.
11 Kitz, *Shattered City*, 58.
12 Kinnear, *In Subordination*, 107.
13 St Martha's Hospital Fund Campaign, "The Story of St Martha's Hospital," 9.
14 Reiser, *Medicine and the Reign of Technology*, 153.
15 Shearer, ed., "Canadian Society of Laboratory Technologists."
16 NSARM MG20, Vol. 197, MHHC, 12 August 1920.
17 MacKenzie, "Origin and Development," 182-3.
18 A.G. Nicholls, "Report of the Work of the Public Health Laboratory" in PHAR, October 1922–September 1923, 21, October 1923–September 1924, 30.
19 PANB RS 136, Records of the Deputy Minister of Health, File E5a, Roberts to Walsh, 1 October 1919.
20 Sandelowski, *Devices and Desires*, 1.
21 Of course, similar processes unfolded in office work and other areas. See, for example, Lowe, *The Administrative Revolution*.
22 Reverby, *Ordered To Care*, 114.
23 Sandelowski, *Devices and Desires*, 83.
24 NSARM, RG25, Series B, Vol. 3, BOC, 16 May 1914; and ibid., Vol. 2, MMB, 24 November 1919. Eager stayed in this position until 1926; see ibid., BOC, 13 August 1926.
25 VGHL, W.W. Kenney to W.H. Eagar, 10 November 1919. In correspondence with the Honourable E.H. Armstrong, the minister of Public Works and Mines, Kenney wrote that the Victoria General "will supply a *technician* [my emphasis] to this department." VGHL, Kenney to Armstrong, 26 November 1919.
26 VGHL, W.W. Kenney to Mary Noonan, 20 January 1920.
27 VGHL, W.W. Kenney to Dr A.F. Miller, Superintendent, Nova Scotia Sanatorium, 27 January 1920. While unnamed, it is likely that the man was Michael MacInnis, who would serve as X-ray technician well into the 1940s. See NSARM, RG25, Series B, Vol. 3, BOC, 2 June 1943.
28 PHAR, October 1922–September 1923, 21.
29 PHAR, October 1923–September 1924, 30.
30 Dhirendra Verma suggested that "from 1914 to 1922, young women were trained in the laboratory to conduct a few simple tests. There was no

definite pattern of training." Verma also suggested that "young high school graduates were chosen by hospitals, shown how to perform various tests in the laboratory, and were appointed permanently as regular laboratory technologists." Verma, "Medical Laboratory Technology Instruction," 5, 19. While no evidence could be found to refute or substantiate this assertion, it is a reasonable inference that the longer training course was in all likelihood intended for such young women.
31 VGHL, W.W. Kenney to Dr W. Eagar, 4 November 1924.
32 This material is drawn from NSARM, Series 018, Social History of Nursing in Nova Scotia in the 1930s. The entire collection consists of thirty-four interviews, fifteen of which were collected by Barbara Keddy and nineteen by students in her research methods course. All of the interviews used here were conducted by Keddy. I am grateful to Dr Keddy for her permission to cite from this important collection.
33 NSARM Keddy Collection, Series 18, MF160–11, interview with Greta MacPherson, 16 March 1983. When Keddy asked MacPherson about radiation exposure from doing the x-ray work, MacPherson replied that "I always blamed that for my infertility," suggesting the hazards associated with this work.
34 Ibid., MF160–10, interview with Flora K. McDonald, 6 October 1982.
35 Ibid., MF160–2, interview with Dorothy Allan.
36 Ibid., MF160–4, interview with Clara M. Buffet, 7 October 1982.
37 Ibid., MF160–10, interview with Flora K. McDonald, 6 October 1982.
38 Pothier, *Mary Ann Watson*, 52.
39 Weir, *Survey of Nursing Education in Canada*, 263.
40 *Echo* (Halifax), 20 March 1925.
41 For examples of training in the x-ray department, see VGHL, Kenney to Edmund MacDonald, Secretary, Harbour View Hospital, 29 August 1922, Kenney to J.S. Calder, Superintendent, City of Sydney Hospital, 21 January 1924, and Kenney to Dr J.K. MacLeod, 22 January 1924. A course in hospital pharmacy work was not offered until the early 1930s. See NSARM, RG25, Series B, Vol. 3, BOC, 24 April 1931, 8 May 1931, and 21 May 1931.
42 VGHL, W.W. Kenney to Miss M.E. MacKay, 30 October 1925.
43 VGHL, W.W. Kenney to Florence H. Merlin, 29 November 1924. See also ibid., Kenney to Eagar, 4 November 1924, and Kenney to Merlin, 6 November 1924. The fact that MacDonald arrived unexpectedly arose out of the continuing confusion over who administered the laboratory. Merlin wrote directly to the laboratory director, Nicholls, to arrange for the course. Nicholls requested that she write to Kenney. In the meantime, the laboratory director received a telegram saying that MacDonald was on her way. Kenney accommodated the nurse but noted that the laboratory or x-ray departments could not "undertake any special or extraordinary obligations without the knowledge and consent of the executive of this hospital."

44 VGHL, W.W. Kenney to Dr W. Eagar, 4 November 1924. In another instance Kenney wrote in early 1924 that the hospital "felt for a long while that it was perhaps one of our public duties to assist in any way we possibly could the smaller hospitals of the province." VGHL, Kenney to Miss J.S. Calder, Superintendent, City of Sydney Hospital, 21 January 1924.
45 VGHL, Kenney to A.G. Nicholls, 17 November 1925.
46 NSARM, RG25, Series B, Vol. 3, BOC, 12 November 1925.
47 VGHL, Kenney to Eagar, 13 November 1925, and NSARM, RG25, Series B, Vol. 3, BOC, 25 November 1925. Eagar negotiated a fee of thirty dollars for his part, but Nicholls requested a larger sum. Correspondence does not record what amount was ultimately negotiated by Nicholls. See VGHL, Kenney to Miss M.E. MacKay, 16 November 1925, Kenney to Eagar, 19 November 1925, Kenney to Eager, 26 November 1925, and Kenney to Nicholls, 1 December 1925.
48 Sandelowski, *Devices and Desires*, 3.
49 Ibid., 85.
50 Ibid., 63–4, 72.
51 Stevens, *In Sickness and in Wealth*, 12.
52 Reverby, "Neither for the Drawing Room," 260; Sandelowski, *Devices and Desires*, 83–6.
53 On the MHHC, see Buxton, "Private Wealth and Public Health," 183–93.
54 NSARM, RG25, Series B, Vol. 3, BOC, 25 May 1921. Nicholls was, of course, aware of the MHHC's desire to expand the staff. In January the commission discussed appointing an assistant to Nicholls who would also be able to devote his entire time "to the work for this Commission." At a subsequent meeting held on 12 August 1920 the following resolution was passed: "Resolved that the [MHHC] have learned with interest that it is the intention of the Provincial Government to extend the Provincial Pathological and Public Health Laboratory in this City and this Commission begs leave to press upon the Government the necessity of proceeding with the work at the earliest possible moment." The minutes go on to note that the increased work "will necessitate a full time assistant and also an additional technician in the Laboratory, and Drs Nicholls, Hattie and Royer and looking for suitable persons but up to the present have not been able to locate them." NSARM, MG20, Vol. 197, Minutes of the MHHC, 17 January, 26 January, and 12 August 1920.
55 PHAR, October 1920–September 1921, 23, and NSARM, MG20, Vol. 197, Minutes of the MHHC, 7 April, 21 April, and 19 May 1921. Morse, who came to Halifax from the Montreal General Hospital, was appointed as the assistant to Dr Nicholls at a salary of $3,000 a year.
56 PHAR, October 1921–September 1922, 23, and NSARM, MG20, Vol. 197, Minutes of the MHHC, 22 March 1922.
57 NSARM, RG25, Series B, Vol. 3, BOC, 31 May 1923 and 25 October 1923; VGHL, W.W. Kenney to G. Fred Pearson, 6 November 1923; NSARM,

MG20, Vol. 197, Minutes of the MHHC, 17 October 1923. Murray was appointed at a salary of $2,000 a year. Before assuming his position, he went to Johns Hopkins School of Hygiene to take a nine-week course, for which the commission advanced him $400. This amount was repaid by Murray in monthly instalments, however. The investment was sound.
58 JHA 1922.
59 NSARM MG20, Vol. 197, Minutes of the MHHC, 2 June 1921, and 30 August 1922.
60 Ibid., 8 August 1923, and PHAR, 1922–23.
61 NSARM, RG25, Series B, Vol. 3, BOC, 31 May 1923.
62 DUAPO, A-817, A.G. Nicholls to A.S. MacKenzie, 9 May 1923, and NSARM, RG25, Series B, Vol. 3, BOC, 31 May 1923; *Nova Scotia Medical Bulletin* 5, no. 11 (November 1926): 36, and vol. 6, no. 9 (September 1927): 35. See also MacLeod, *Petticoat Doctors*, 90–1. The entry for Chase records nothing of her work in the laboratory, however.
63 This account is based upon MacLeod, *Petticoat Doctors*, 9–91.
64 Murray, incidentally, worked as a demonstrator in anatomy for a brief period before leaving for Korea. See Brouwer, "Beyond 'Women's Work for Women,'" 65–95.
65 Kinnear, *In Subordination*, 61.
66 Morantz-Sanchez, *Sympathy and Science*, 61–2.
67 Women's exclusion and constrained opportunity in areas such as medicine, science, and university teaching prepared the ground for women's entry into allied health care work. In this way, there are parallels between women's entry into health care work and women's exclusion from skilled work, at least at the theoretical level. For relevant studies of women and skilled work, see Cockburn, *Brothers*, and Rose, *Limited Livelihoods*, especially chapters 2 and 6.
68 NSARM, RG25, Series B, Vol. 3, BOC, [?] August 1924, and VGHL, W.W. Kenney to A.G. Nicholls, 3 December 1924, and Kenney to Nicholls, 30 December 1924.
69 NSARM, RG25, Series B, Vol. 3, BOC, 23 September 1925.
70 Ibid., and VGHL, W.W. Kenney to A.G. Nicholls, 24 September 1925.
71 NSARM, RG25, Series B, Vol. 3, BOC, [?] August 1924 and 29 December 1924. VGHL, W.W. Kenney to A.G. Nicholls, 30 December 1924.
72 DUAPO, A-817, "Memo re: Staff Pathological Department. Victoria General Hospital," October 1926.
73 NSARM, RG25, Series B, Vol. 3, BOC, 25 November 1925.
74 DUAPO, A-817, A.G. Nicholls to A.S. MacKenzie, 3 September 1924.
75 DUAPO, A-817, A.G. Nicholls to A.S. MacKenzie, 3 September 1924. Robinson later registered as a student in histology, and Nicholls again exacted a tuition exemption from Dalhousie, apparently in exchange for her work as a laboratory assistant in bacteriology. Robinson, Nicholls wrote, "will be

very useful in this particular work and will well earn the amount in question." See ibid., A.G. Nicholls to Prof. Murray MacNeill, 15 September 1926.
76 DUAPO, A-817, "Memo re: Staff Pathological Department. Victoria General Hospital," October 1926.
77 NSARM, RG25, Series B, Vol. 3, BOC, 12 February 1931.
78 Ibid., 25 November 1925.
79 DUAPO, A-817, "Memo re: Staff Pathological Department. Victoria General Hospital," October 1926. In 1930 Pathology director Ralph P. Smith requested a full-time assistant; the request was turned down by the medical board of the hospital. Dalhousie apparently wanted Smith to devote more time to the medical school, a stance the hospital rejected outright. The medical board explained that "the Hospital would have the first claim on the Pathologist's time, and that the University would accept what time the Pathologist had to give after he had rendered full service to the position of Provincial Pathologist, and the Board was of the opinion that if this understanding was fully implemented there would be no necessity for a full time assistant." NSARM, RG25, Series B, Vol. 3, BOC, 24 February 1930.
80 PHAR, October 1927–September 1928, 21.
81 DUA, Dalhousie University Student Registers.
82 PHAR, October 1929–September 1930, 28.
83 PHAR, December 1937–November 1938, 35.
84 PHAR, October 1933–September 1934, 23, and October 1934–November 1935, 22.
85 Interview with Rose Phillips, conducted by Peter L. Twohig, 22 April 1996.
86 NSARM, RG25, Series B, Vol. 3, BOC, 18 January 1933.
87 Interview with Edna Williams, conducted by Peter L. Twohig, 23 April 1996.
88 Interview with Rose Phillips, conducted by Peter L. Twohig, 22 April 1996.
89 It is not clear from the reference whether the assistants and learners were the same people, or whether there was a distinction being made between the two. Certainly, with the training that was going on in the Saint John facility in this period, learners could also have been unpaid assistants. See NBARMH, 31 October 1933, 10.
90 NBARMH, Annual Report of the Bureau of Laboratories Year Ending 31 October 1936, 4.
91 CSLT, Membership Files.
92 NBARMH, Annual Report of the Bureau of Laboratories Year Ending 31 October 1934, 4. The report notes the addition of several additional volunteers, while one of the previous year's volunteers left to assume a position at Amherst's Highland View Hospital.
93 Ibid., Year Ending 31 October 1937, 4. Two "apprentices" from the previous year, Constance Fewings and Frances Crocker, were added to the staff of the Saint John facility in 1936–37.

94 New Brunswick, *Public Accounts*. Crocker resigned in 1938–39 to return to her native Fredericton. See NBARMH, Annual Report of the Bureau of Laboratories Year Ending 31 October 1939, 4.
95 NBARMH, Annual Report of the Bureau of Laboratories Year Ending 31 October 1936, 4. The annual report for 1935–36 clearly identifies Hunter as a "voluntary member of the staff" for the two previous years, but the public accounts reveal that she did, in fact, earn some income the previous year, the small sum of $118.68 in 1934–35. Hunter was initially paid for "lab services," then named a "lab assistant" before becoming the "senior technician." See New Brunswick, *Public Accounts*.
96 PANB RS 136, File A1a, "The New Brunswick Department of Health," 2 June 1930.
97 NBARMH, 31 October 1933, 10.
98 NBARMH, 31 October 1933, 10.
99 PANB RS 136, Records of the Deputy Minister of Health, File E5a, William F. Roberts to Sister Superior Walsh, 1 October 1919.
100 New Brunswick, *Public Accounts* recorded that Marks earned $1,000 in 1919–20 and $1,200 in 1920–21, which was increased to $1,350 in 1923–24 and $1,500 in 1924–25, the figure she earned until retirement. See also NBARMH, 31 October 1929, 5.
101 *Pharos 1930*, 35.
102 CSLT, Membership Files.
103 New Brunswick, *Public Accounts* 1934–35.
104 NBARMH, Annual Report of the Bureau of Laboratories, 31 October 1935, 4.
105 New Brunswick, *Public Accounts*. This figure is suggestive, because it works out to fifty dollars per month. There is every possibility that he was paid the same amount in the previous three years but only worked for one, six, and five months, respectively. His salary increased to $660 the following year, but beginning in 1939–40 the sums diminished greatly, and it would appear that Stewart performed only selected tests for the Bureau of Laboratories.
106 NBARMH, Annual Report of the Bureau of Laboratories, 31 October 1935, 4.
107 New Brunswick, *Public Accounts*.
108 In addition to women who performed tests, married women also worked to clean the laboratory, including a Mrs Night in the late 1930s and Mrs Angeline Gaugin in the early 1940s. See ibid., *Public Accounts*.
109 CSLT, Membership Files.
110 See New Brunswick, *Public Accounts* and NBARMH, Annual Report of the Bureau of Laboratories Year Ending 31 October 1936, 4. Incidentally, when she resigned, Mullins was replaced by Sybil Pelton, who had served as a volunteer for two years previously.

111 A.R. Shearer to Gloria D. Gould, 12 July 1973, and Shearer to McGeouch, 7 August 1973 in CSLT Correspondence Files.
112 New Brunswick, *Public Accounts*.
113 NBARMH, Annual Report of the Bureau of Laboratories Year Ending 31 October 1942, 3–4.
114 Burns described himself as "Chief Technician" in his application to the CSLT. This designation does not appear in the public accounts. Burns was described variously as the laboratory assistant and paid for laboratory services. Nevertheless, he was the highest paid worker in the laboratory until his departure, which bolsters his claim.
115 CSLT, Membership Files.
116 New Brunswick, *Public Accounts*, and NBARMH, Annual Report of the Bureau of Laboratories Year Ending 31 October 1939, 4.
117 New Brunswick, *Public Accounts*, and NBARMH, Annual Report of the Bureau of Laboratories Year Ending 31 October 1942, 3. On the difficulty of maintaining an adequate staffing complement during the war, see NBARMH, Annual Report of the Bureau of Laboratories Year Ending 31 October 1941, 3.
118 DUAPO, A-606, B[oris] Babkin to A. Stanley MacKenzie, 9 September 1925; Babkin to MacKenzie, 28 September 1925; Miss Harris to Babkin, 6 January 1926; Babkin to Harris, 30 April 1926.
119 Ibid., Babkin to MacKenzie, 4 September 1926; Babkin to MacKenzie, 11 September 1926; Babkin to Miss Harris, 27 September 1926; Babkin to MacKenzie, 1 May 1927; Babkin to MacKenzie, 15 September 1927; E.W.H. Cruickshank to MacKenzie, 13 March 1929; Cruickshank to MacKenzie, 18 April 1929.
120 Ibid., Cruickshank to MacKenzie, 18 April 1929, and MacKenzie to Cruickshank, 29 April 1929.
121 Ibid., C.W. Startup to Miss Harper, 3 July 1929; Startup to Harper, 19 August 1929.
122 Information on all these individuals may be found in DUAPO, A-606.
123 DUAPO, A-597, E. Gordon Young to A.S. MacKenzie, 3 October 1923, and MacKenzie to Young, 9 November 1923. The emphasis on gender, which is mine, is significant given the evidence that men were employed in the biochemistry lab through the 1920s and 1930s.
124 DUAPO, A-597, Young to MacKenzie, 28 November 1923.
125 Ibid., Young to Professor H.R. Theakston, 24 February 1926; Young to MacKenzie, 26 January 1927; MacKenzie to Young, 2 March 1927; Young to MacKenzie, 14 September 1927; Young to MacKenzie, 28 October 1928; Young to H.L. Harper, 28 June 1929.
126 Ibid., Young to Harper, 12 September 1929; Young to Harper, 28 October 1929; MacKenzie to Young, 1 November 1929; Young to MacKenzie, 12 November 1929; MacKenzie to Young, 14 April 1930; Young to

Miss H. Joyce Harris, 25 September 1930; Young to MacKenzie, 18 March 1931; MacKenzie to Young, 24 March 1931. In her analysis of marriage in north-end Halifax, Suzanne Morton used a small sample of twenty-four marriages in 1921 to note that working-class people from that neighbourhood may have married at a younger age. The average age for marriage was 25.6 years for men, below the national average of twenty-eight. See Morton, "The June Bride as the Working-Class Bride," 367.
127 DUAPO, A-597, Young to Carleton W. Stanley, 6 October 1931.
128 Ibid., Young to H.G. Grant, 2 November 1939.
129 DUAPO, A-606, see E.W.H. Cruickshank to Carleton Stanley, 8 December 1931; Stanley to Cruickshank, 11 December 1931; Cruickshank to Stanley, 21 March 1932; Stanley to Cruickshank, 22 March 1932; Cruickshank to Stanley, 4 October 1932; H.G. Grant to Stanley, 3 June 1937.
130 Ibid., E.W.H. Cruickshank to the President, 8 December 1931, and Carleton Stanley to Cruickshank, 11 December 1931.
131 Ibid., Cruickshank to President, 4 October 1932. Howitt's eldest son, incidentally, was also employed in the medical school, working for Dr N.B. Dreyer of the physiology department.
132 In 1931–32, the medical school expended almost $1,100 for assistants in anatomy, $750 for biochemistry, $1,130 for pathology, $570 for pharmacy, and $573 for physiology. See DUAPO, A-575, "Analysis of Medical School Costs," 30 June 1932. Occasionally, "technical assistants" were supported through grants. Donald Maitland of the department of anatomy received a grant from the Banting Research Foundation and he chose to employ his wife, who had experience measuring the size of nuclei and cells. See DUAPO, A-595, Donald Maitland to Carleton Stanley, 15 November 1931.
133 Montgomery, *The Fall of the House of Labor*, 239, and also *Workers' Control in America*, 32–3.
134 Sangster, *Earning Respect*, 74. While the importance of family wage ideology is unquestioned, wages were, as Bettina Bradbury has suggested, "seldom the only source of survival" for families. See Bradbury, *Working Families*, 48.

CHAPTER FIVE

1 Walker, "The Technician's Trials," 71.
2 CSLT, Membership Files, M.J.D. to A.R. Shearer, 20 September 1961. The range of employees also represented a substantial expansion of the workforce following the Second World War. This worker completed her laboratory training in 1943 and then returned to complete her baccalaureate education at the University of New Brunswick. She graduated in May 1944 and joined the federal civil service the following month. When she assumed her position at Lancaster, there was only herself and one other technician.

3 New Brunswick, *Public Accounts*, and NBARMH, Annual Report of the Bureau of Laboratories 31 October 1942, 4.
4 New Brunswick, *Public Accounts*, and NBARMH, Annual Report of the Bureau of Laboratories 31 October 1944, 3.
5 PHAR, December 1937–November 1938, 35.
6 NSARM, RG25, Series B, Section VI.16.
7 NSARM, RG25, Series B, Vol. 3, BOC, 7 May 1927.
8 New Brunswick, *Public Accounts*.
9 Interviews with Ellen Robinson, 29 April 1996, and Rose Phillips, 22 April 1996.
10 CSLT, Registry File.
11 *Canadian Hospital* 7 (December 1930): 35.
12 CSLT, Registry File.
13 *Annual Report of the Miramichi Hospital*, 1 May 1930 to 30 April 1931.
14 Thorngate, *That Far Horizon*, 31.
15 *Annual Report of the Directors of the Chipman Memorial Hospital*, years ending 31 December 1921 and 1922.
16 NSARM, Keddy Collection, Series 018, MF 160–13.
17 Ibid., MF 160–27. The emphasis is mine.
18 Sandelowski, *Devices and Desires*, 63.
19 In the United States by 1923, there were over 1,700 nursing schools. See Reverby, *Ordered to Care*, 61. Of course, given such numbers, there was tremendous variation among the schools. Reverby notes that in Boston, hospitals with as few as sixteen beds or as many as one thousand had nursing schools. The growth of nursing schools in Canada was also substantial in the early decades of the twentieth century. In 1909 there were seventy schools of nursing in Canada and this expanded to over two hundred by the 1920s. In 1931 there were 256 schools of nursing across Canada, over half of which were in Ontario (ninety-eight) and Quebec (forty-six). Nova Scotia had eighteen schools, New Brunswick had sixteen, while tiny Prince Edward Island boasted three schools. See McPherson, *Bedside Matters*, 30, and "Skilled Service and Women's Work," 175. Most of these schools required three years of training, so the contribution of "student nurses" to the labour supply for the hospital was substantial. By the later 1930s, any hospital in Nova Scotia with over twenty beds could operate a nursing school, effectively ensuring a labour supply for even small rural hospitals. Twohig, *Challenge and Change*, 11–12.
20 NSARM, Keddy Collection, Series 018, MF 160–35.
21 Reverby, *Ordered to Care*, 27; Keddy, "Private Duty Nursing Days," 99–102.
22 McPherson, "Nurses and Nursing," 26–7.
23 Weir, *Survey of Nursing Education in Canada*, 301. The Canadian Medical Association and Canadian Nurses' Association jointly sponsored this

inquiry into nursing education across Canada. Weir was a professor of education at the University of British Columbia who would become education minister and assume responsibility for the health portfolio in BC under the Liberal government.

24 The CSLT graciously allowed me to see these files. I would like to thank Lynn Zehr and Kurt Davis for facilitating this access. The arrangement does not permit me to identify individuals by name, but this section is based upon a reading of all members of the Maritimes from 1937 to 1945.

25 Axelrod, *Making a Middle Class*, 90.

26 It is not my intention to relate the history of the CSLT here. Readers interested in the details of the professional society will find a fuller account in Twohig, "Organizing the Bench," ch. 5, or Shearer, ed., "Canadian Society of Laboratory Technologists."

27 CSLT, Minutes of the Executive, 8 November 1936. The other objectives, contained in the bylaws, were to promote research, to promote cooperation between laboratory workers and the medical profession, and to "more efficiently" aid in the diagnosis and treatment of disease.

28 Strauss with Bucher, "Professions in Process," 10. This essay was originally published in the *American Journal of Sociology* 66 (January 1961): 325–34.

29 Shearer, ed., "Canadian Society of Laboratory Technologists," 3.

30 "Report of the First Annual General Meeting," *CJMT* 1 (1938): 24. Ninety-seven members came from Ontario, although every other province was represented as well, as was Newfoundland. Membership was as follows: ninety-seven from Ontario, thirty-four from British Columbia, fifteen from Nova Scotia, fourteen from Saskatchewan, eight each from Alberta and Quebec, seven from Manitoba, six from New Brunswick, one from Prince Edward Island, and three from Newfoundland.

31 CSLT, Membership Files, Ronald Burns to CSLT, 11 February 1937.

32 See Thorngate, *That Far Horizon*, 8; Brown, "The Division of Laborers," 76–8; and Shearer, "Canadian Society of Laboratory Technologists," 1. See also Ikeda, "Survey of Training Schools for Laboratory Technicians," 467–76.

33 NBARMH, Annual Report of the Bureau of Laboratories, 31 October 1936, 4.

34 Ibid. That so many of the laboratory workers in Saint John successfully completed the exams and were registered with the ASCP was taken as evidence of the "very excellent training" provided through the Bureau of Laboratories.

35 NBARMH, Annual Report of the Bureau of Laboratories, 31 October 1938, 4. That year Cathy Arnold joined the Canadian society, while a "student technician," Evelyn Russell, gained registration with the ASCP.

36 Interview with Rose Phillips, conducted by Peter L. Twohig, 22 April 1996.

37 CSLT, Membership Files. No systematic analysis was undertaken for other regions of Canada. Among the organizers of the CSLT, however, eight of ten of the charter members held ASCP "MT" designations.
38 CSLT, Membership Files.
39 Sandelowski, *Devices and Desires*, 13.
40 *Canadian Hospital* 7 (October 1930).
41 Ibid., vol. 8 (March 1931) and (August 1931).
42 *Canadian Nurse* 33 (November 1937): 566.
43 Ibid., vol. 34 (July 1938): 378.
44 Ibid., vol. 37 (December 1941): 856.
45 *Canadian Hospital* 9 (February 1932). Although she did not claim any experience in the field, Dobson's advertisement also states that she "Understands x-ray technique," perhaps an indication of her ability to learn this area, if it was required.
46 Ibid. (December 1942).
47 See Ibid., 7 (December 1930): 35.
48 "How Large a Hospital Should Employ a Dietitian-Laboratory Worker?" Ibid., vol. 18 (December 1941): 36.
49 Katz, "The Emergence of Bureaucracy in Urban Education," 155–87.
50 Reverby, "The Search for a Hospital Yardstick," 206.
51 The issue of maintaining discrete work boundaries has been explored most thoroughly for the case of nursing. Barbara Melosh has suggested that in the United States, both professionalizers and rank-and-file nurses worried about the role of nursing auxiliaries, fearing they would "undercut the private-duty market and threaten the place of the graduate nurse." However, after about 1940, nursing leaders began to play an active role in the development of practical nursing, while those in the rank and file retained their suspicion of the auxiliaries. "Essentially," Melosh concluded, "leaders accepted the redefinition of nursing as a middle-management function, while ordinary nurses remained wedded to older craft values, which placed bedside nursing at the center of their work." Melosh, "The Physician's Hand," 177–8.
52 On the multitasking of nurses, see McPherson, *Bedside Matters*, especially 95–107.
53 Ibid., 21. McPherson points out elsewhere that student nurses did not just perform less desirable (i.e., "subsidiary") tasks. Rather, they performed duties that would come to be assigned to physiotherapists, pharmacists, and other allied health care workers, who would not become a regular feature of patient care until after World War II. See McPherson, "Skilled Service and Women's Work," 5.
54 McPherson and Stuart, "Writing Nursing History in Canada," 5–6.
55 McPherson, *Bedside Matters*, 220–1. It bears pointing out that the meaning of nursing practice varied considerably in space and time, and that the

"nurse" is not an essentialist category easily understood across decades and in different places. New tasks were frequently assigned to nurses or old ones withdrawn. As early as 1915, for example, the Victoria General Hospital approved the training of nurses in anesthesia "with the view to employing them as anesthetists." See NSARM, RG25, Series B, Vol. 3, BOC, 2 January 1915. However, as hospitals expanded, the question of boundary maintenance gained importance.

56 This is an ongoing debate among health care professions. For a recent example from nursing, see Cooke, "Boundary Work in the Nursing Curriculum," 1,990–8. The abstract for this article suggests that "boundary work enables a discipline to stake out a claim to its legitimate territory and the resources that go with it. In a practice discipline such as nursing, the boundaries between nursing and supporting subjects, such as sociology and physiology, create problems of transfer of learning."

57 Charles and Fahmy-Eid, "La Diététique et la Physiothérapie Face au Problème des Frontières Interprofessionnelles (1950–1980)."

58 Sarah Jane Growe, *Who Cares?*, 99–100.

59 Wright, "Administration in Small Hospitals," 230. Kathryn McPherson suggests that for student nurses, too, a variety of tasks were recounted in student yearbooks, including work in laboratories. See McPherson, *Bedside Matters*, 109.

60 NSARM, Keddy Collection, Series 018, MF 160–9. This interview is erroneously listed as Grace Long in the finding aid to the Keddy Collection.

61 Morrison, "The Nurses in Hospital Administration," 672–4.

62 Sister Catherine Gerard, "We Look at Nursing Service," 827.

63 Besides the two cases discussed, there were other circumventions of the admission rules in Saint John. In 1947 two students were admitted to the training program having completed only one science credit, instead of the mandatory two. The CSLT decided to let these women sit for the national exams because the change to the admission policy had been a recent one and "Dr MacKeen accepted these girls not knowing of our regulations at that time." Moreover, the society was confident that "it would not occur again." See CSLT, Minutes of the Executive, 18 October 1947.

64 CSLT, Correspondence Files, H.B. to Ileen Kemp, 29 June 1945.

65 CSLT, Membership Files.

66 Ibid., and CSLT, Correspondence Files, and F.W. Patterson to Ileen Kemp, 21 September 1943. The student completed courses in English, two courses in household economics, and a half-credit in botany in 1941–42. She did not, however, successfully complete a course in chemistry or a zoology half-credit.

67 CSLT, Correspondence Files, Ileen Kemp to M.D., 24 February 1944.

68 Ibid., M.R. to Ileen Kemp, 29 June 1945.

69 Ibid.

70 Elliot, "Open Letter," 6.
71 Ibid., 6.
72 CSLT, Minutes of the Executive, 8 November 1936.
73 CSLT, Minutes of AGM, 11 December 1937.
74 Interview with Edna Williams, conducted by Peter L. Twohig, 23 April 1996.
75 Ibid.
76 Indeed, in 1937–39, the formative years of the CSLT, Father George Verreault headed the Canadian Hospital Council. See G. Harvey Agnew, *Canadian Hospitals, 1920 to 1970*, Appendix C. The national Catholic Hospital Council of Canada was not founded until the early 1940s.
77 Newfoundland was included in this sample because of the strong connections between it and the Maritime provinces. Many women from Newfoundland trained in Maritime hospitals through the 1920s, 1930s, and 1940s.
78 See for example CSLT, Membership Files, J.R.W. to D.R. Lock, 12 March 1937.
79 *Canadian Journal of Medical Technology* 1 (October 1938): 20.
80 Ibid. (September 1939): 149. This edition was only a modest thirty pages, perhaps indicating the difficulty of producing a journal entirely on volunteer submissions and labour over the summer months. The prevalent voluntarism at the national office is discussed below. Before the *CJMT* began publication, Denys Lock sent a letter to the membership informing them of the journal and asking them to "please consider this letter as a personal appeal to you and send in a contribution in the form of an article on some technique which you yourself may have originated." CSLT, Correspondence Files, D.R. Lock to Member, 25 June 1938. The request for this particular kind of submission is suggestive of the shape the executive wished the journal to assume, namely, that of a scientific journal rather than a newsletter or other such publication.
81 CSLT, Minutes of AGM, 31 May 1940.
82 Ibid.
83 Ibid., 6 June 1955.
84 Ibid.
85 CSLT, Minutes of the Executive, 28 May 1947.
86 Ibid., 21 January 1951.
87 Ibid., 1 April 1951.
88 CSLT, Membership Files, September 1945.
89 CSLT, Minutes of the Executive, 6 January 1937.
90 Shearer, "Canadian Society of Laboratory Technologists," Appendix D.
91 CSLT, Minutes of AGM, 20 May 1949, 26 June 1950, and 25 June 1951.
92 Ibid., 20 May 1949.
93 Ibid., 26 June 1950. When the Manitoba organization entered into discussions with the CSLT, there was a question of membership. Some of the

Manitoba members were not members of the national society, although they were "most actively interested in their Manitoba Society." The Manitoba executive was keen to maintain these members and asked the national office whether they could be admitted. The CSLT decided that they could be, as "auxiliary" members, without voting privileges and without being eligible for office. See CSLT, Minutes of the Executive, 25 June 1950. In Nova Scotia, Sister Agnes Gerard worked tirelessly to form a Nova Scotian branch, but to no avail. See ibid., 1 April 1950.
94 CSLT, Minutes of AGM, 25 June 1951.
95 Ibid., 7 June 1954; CSLT, Executive Meeting Detailed Agenda, 20–21 March 1954; and Shearer, "Canadian Society of Laboratory Technologists," Appendix D.
96 CSLT, Membership Files.
97 Ibid.
98 CSLT, Minutes of the Executive, 26 July 1947.
99 Ibid., Minutes of AGM, 1 June 1946.
100 Ibid., Minutes of the Executive, 2 June 1946.
101 NBARMH, Annual Report of the Bureau of Laboratories, 31 October 1941, 3.
102 Ibid., 31 October 1945, 4.
103 CSLT, Membership Files.
104 Ibid.
105 Ibid., Minutes of the Executive, 31 January 1948. See also ibid., 10 April 1949. It is unclear where the complaint originated, but it appears to have come from both the students and the laboratory workers.
106 Ibid., 22 January 1950.
107 "Saskatchewan Hospitals Protest," 41.
108 Canada, Department of Labour, *Medical Laboratory Technologist*.
109 Ibid.
110 On health grants, see CSLT, Minutes of the Executive, 1 April 1950; Taylor, *Health Insurance and Canadian Public Policy*, 162–4; and Naylor, *Private Practice*, 132–4, 153. For discussion of fees for training, see CSLT, Minutes of the Executive, 18 May 1952. The only direct evidence of tuition charges in the Maritimes was an application from Saint John that indicates that the student paid twenty-five dollars to enroll in the thirteen-month course. See CSLT, Membership Files. Indirect evidence from Saint John also suggests that several years earlier, the Bureau of Laboratories considered the work students performed more than sufficient. The laboratory report for 1931–32 noted that "several young ladies and two young men" were taking instruction in laboratory technique and that "they render help in return for the instruction which they receive."
111 CSLT, Minutes of the Executive, 26 July 1947.
112 Ibid., 18 October 1947.

113 *Canadian Nurse* 41 (July 1945): 552. Two schools in Quebec offered postgraduate courses in x-ray technology, the Phillips Training School for Nurses at Montreal's Homeopathic Hospital and the Hôpital Générale Saint-Vincent de Paul in Sherbrooke.
114 *Canadian Nurse* 37 (January 1941): 34.
115 NAC, MG28–I343, CMA, Minutes of Executive Committee, 18–19 June 1937; *Canadian Journal of Medical Technology*, 1 (October 1938): 5; Shearer, "Canadian Society of Laboratory Technologists," 2–3. The other members were Dr E.H. Mason, Montreal, Dr James Miller, Kingston, Dr George Shanks, Toronto, Dr J.C. Patterson, Regina, and Dr J.J. Ower, Edmonton.
116 CSLT, Minutes of AGM, 11 December 1937.
117 "Report of the Committee on Laboratory Technicians," NAC, MG28–I343, CMA, Minutes of General Council, 19–20 June 1939; CSLT, Minutes of the Executive, 8 February 1941.
118 NAC, MG28–I343, CMA, Minutes of Executive Committee, 20–21 June 1941. See also ibid., 14–15 March 1941.
119 *Canadian Journal of Medical Technology* (March 1942). By the next CMA General Council meeting, there were ten approved schools and the applications of several other laboratories were pending. NAC, MG28–I343, CMA, Minutes of General Council, 15–16 June 1942.
120 The approved schools were in Halifax, Saint John, Montreal, Ottawa, Kingston, two in Toronto (St Michael's Hospital and Toronto Western Hospital), and two in Hamilton (Hamilton General Hospital and Mountain Sanatorium).
121 CSLT, Correspondence Files, A.D. Kelley to Helen L. Smith, 3 November 1948.
122 NAC, MG28–I343, CMA, Minutes of Executive Committee, 30–31 October 1941; ibid., Minutes of General Council, 15–16 June 1942.
123 CSLT, Correspondence Files, A.D. Kelly to Helen Smith, 3 November 1948.
124 Ibid.
125 CSLT, Minutes of the Executive, 27 May 1951.
126 Ibid., Minutes of AGM, 25 June 1951.
127 CSLT, Minutes of the Executive, 6 November 1948. This was a minority opinion but one that originated in the CMA Council.
128 References to university education were frequent in the CSLT documents. See for example Minutes of the Executive, 27 January 1940 and 11 March 1944.
129 CSLT, Correspondence Files, Robert T. Noble to W.J. Deadman, 29 March 1948.
130 O'Donnell, "O Pity the Poor Student," 41.
131 CSLT, Minutes of AGM, 1 June 1946.

132 The CSLT recognized that "the need in small hospitals is very great for technicians who can do blood groupings, blood counts and urinalysis and x-rays of chests and fractures." The frequent combination of these skills led the Society to discuss affiliation with the radiology society in the mid-1940s. See Minutes of AGM, 1946.
133 *Canadian Hospital* 23, 11 (November 1946): 84.
134 Ibid.
135 CSLT, Minutes of the Executive, 2 June 1946 and 5 October 1946.
136 *Canadian Hospital* 23, 11 (November 1946): 84.
137 Canada, Department of Labour, *Medical Laboratory Technologist*; CSLT, Minutes of the Executive, 19-20 March 1955. It is worth noting that the Saskatchewan government reinstituted the combined x-ray and laboratory course in October 1953. See Shearer, "Canadian Society of Laboratory Technologists," 16.
138 Shearer, "Canadian Society of Laboratory Technologists," 13.
139 CSLT, Minutes of the Executive, 17 May 1952, and ibid., Minutes of AGM, 19 May 1952. Of course, the University of Saskatchewan was by this time also offering an education option for persons who wanted to work in laboratories. The Regina College program garnered a student partial credit toward fulfillment of the university degree requirements.
140 CSLT, Minutes of the Executive, 27 January 1940. The decision of the CMA was in response to inquiries from the University of Western Ontario. See CSLT, Correspondence Files, Denys R. Lock to Harvey Agnew, 12 December 1939, and Agnew to Lock, 18 January 1940. Agnew believed that the rejection was predicated in part on the belief that the "full university course was being considered as the only type of training for technicians."
141 CSLT, Minutes of the Executive, 11 March 1944.
142 Ibid., 10-11 December 1949.
143 NAC, MG28-I343, CMA, Minutes of General Council, "Report of the Committee on Laboratory Technologists," 14-15 June 1943. See also the laboratory committee's report to CMA General Council, 11-12 June 1945.
144 CSLT, Minutes of the Executive, 17 April 1943. The minutes reproduce a letter from W.S. Linday, Dean of Medical Sciences at the University of Saskatchewan.
145 Ibid.
146 Ibid., 31 January 1948. There were differences in the two programs. While students at the University of Saskatchewan took their practical training following three university years, students at the University of Western Ontario took their practical rotations at approved hospitals during their university years.
147 Ibid., 16 December 1951.
148 Ibid., and CSLT, Minutes of AGM, 19 May 1952.

149 These questions were raised specifically in reference to the University of British Columbia program but were generally applied across the country. See CSLT, Executive Meeting Detailed Agenda, 7 December 1952.
150 CSLT, Minutes of the Executive, 21 October 1951.
151 Gerard and Kemp, "Medical Laboratory Technologist."
152 Taylor, *Health Insurance*, 162–4, and Naylor, *Private Practice*, 132–4, 153. National Health and Welfare minister Paul Martin addressed the issue of federal aid to nurse training in *Canadian Nurse* 47 (May 1951): 327–30. On the school's first decade at Dalhousie, see Twohig, *Challenge and Change*, 22–47.
153 Interview with Electa MacLennan by Lynn Kirkwood, 10 December 1983. I am most grateful to Dr Kirkwood for allowing me to use this transcript and to John Gordon for providing me with a copy.
154 CSLT, Minutes of the Executive, 11 March 1944.
155 Ibid.
156 Adams, *A Dentist and a Gentleman*, 110–21.
157 CSLT, Minutes of the Executive, 15 April 1945.
158 Ibid., 14 October 1951.
159 NAC, MG28-I343, CMA, Minutes of General Council, 19–20 June 1939, and *Canadian Journal of Medical Technology* 3 (1941): 154.
160 NAC, MG28-I343, CMA, Minutes of General Council, 14–15 June 1943.
161 CSLT, Minutes of AGM, 11 December 1937; ibid., Minutes of the Executive, 28 May 1937 and 31 January 1938; *Canadian Journal of Medical Technology* 1 (October 1938): 5–6. An undated CSLT "Instructions to Local Examiner" described the duties of the laboratory director for the examinations and the preliminary interview. During the interview, in addition to discussing training and previous experience in the laboratory, examiners were to "size up applicant as to personality and appearance. Submit your opinion in writing when returning the papers."
162 NAC, MG28-I343, CMA, Minutes of Executive Committee, 14–15 March 1941; "Report of the Committee on Laboratory Technicians," ibid., Minutes of General Council, 22–23 May 1943.
163 Minutes of the Canadian Medical Association Nucleus Committee on Technicians, 9 February 1944, in CSLT and "Report of the Committee on Laboratory Technicians," in NAC, MG28-I343, CMA, Minutes of Executive Committee, 3–4 March 1944.
164 Adams, *A Dentist and a Gentleman*, 110–4, 117–18; Witz, *Professions and Patriarchy*, especially chs. 2 and 6; Davies, "The Sociology of Professions and the Profession of Gender."

CHAPTER SIX

1 Details on the creation of the PEI department of health are in Baldwin, "Volunteers in Action," 121–47, and *She Answered Every Call*, 181.

2 McCuaig, *The Weariness*, 122. In addition to the government grant, the Red Cross provided $11,800 and the Maritime Tuberculosis Education Committee $4,500.
3 NAC, CTA, MG28, 175, File 48, Certified Copy of Minutes of His Honour the Lieutenant Governor in Council, 23 December 1930. See also Baldwin, "Volunteers in Action," 145-6.
4 NAC, CTA, MG28, 175, File 54. P.A. Creelman to Dr D.J. MacKenzie, n.d. [March 1931]. This letter is a copy of the original.
5 Ibid., Wodehouse to Marion Merry, 12 February 1931.
6 Ibid., Merry to Wodehouse, 22 February 1931.
7 Ibid., Wodehouse to Defries, 24 February 1931.
8 Ibid., Wodehouse to Keeping, 27 February 1931.
9 For further details on how these factors shaped the selection of public health nurses in industrial Cape Breton, see Twohig, "Public Health in Industrial Cape Breton, 1900–1930s."
10 NAC, CTA, MG28, 175, File 54, Stevenson to Creelman, 23 January 1931. Stevenson had previously expressed some interest in securing a position on Prince Edward Island. She had applied for, and apparently been promised, a nursing position at the Provincial Sanatorium. See ibid., Creelman to Stevenson, 20 January 1930.
11 For salary proposals, see ibid., Wodehouse to Keeping, 3 February 1931 and ibid., File 48, Wodehouse to Mr V.H. Smith, Confederation Life Association, 14 July 1931. P.A. Creelman wrote as early as February 3 that "Miss Stevenson might not require the same salary, for the first year at least, as Miss Merry. In view of her lack of training and experience in this work, I do not think she would be entitled to it either." Ibid., File 54, Creelman to Wodehouse, 3 February 1931.
12 Ibid., File 54, Creelman to MacKenzie, n.d. [March 1931].
13 Reid, "Health, Education, Economy," Table 1, 79, and Twohig, "The Rockefellers, The Cape Breton Island Health Unit and Public Health in Nova Scotia."
14 Vincent, *The Rockefeller Foundation*, 5, 31. The failure of this initiative is discussed in Reid, "Health, Education and Economy," 71.
15 See NAC, CTA, MG28, 175, File 54, Creelman to Wodehouse, 29 April 1931, and Wodehouse to Creelman, 30 April 1931.
16 Ibid., File 48, Executive Secretary [R.E. Wodehouse] to Mr V.H. Smith, Confederation Life Association, 14 July 1931.
17 NSARM, RG25, Series B, Vol. 3 (1910–1943), BOC, 18 January 1936.
18 Ibid., 3 August 1939.
19 NBARMH, Annual Report of the Bureau of Laboratories, Year Ending 31 October 1944, 4.
20 The issue was raised at the annual general meeting of the CSLT in 1942. While the issue was discussed, no action was taken, although the secretary

reported that salaries for workers had increased by about twenty percent since the society was founded. CSLT, Minutes of AGM, 13 June 1942.
21 Of course, standards varied widely. In 1938 a worker in the Digby General Hospital who had attended classes at Dalhousie University earned fifty dollars a month, with meals and a room. She had been at work for six months. Across the Bay of Fundy in Saint John that same year, another new worker earned seventy dollars a month.
22 CSLT, Membership Files, and M.C.H. to Helen L. Smith, 16 March 1942.
23 CSLT, Membership Files.
24 NBARMH, Annual Report of the Bureau of Laboratories, Year Ending 31 October 1942, 5.
25 *Canadian Hospital* 19 (October 1942): 44–50 and ibid. (November 1942): 34–40.
26 Urquhart, ed., *Historical Statistics of Canada*, Series B125, 46.
27 Gagan and Gagan, *For Patients of Moderate Means*, 5–6.
28 David Gagan and Rosemary Gagan have clearly illustrated this growth, noting that hospital revenues grew by more than 267 percent between 1942 and 1951, while hospital revenues increased by nearly 275 percent between 1945 and 1954. See *For Patients of Moderate Means*, 89.
29 CSLT, Minutes of AGM, 29 May 1943.
30 Ibid., 20 May 1944.
31 NAC, MG28–F343, CMA, Minutes of General Council, Montreal, 11–12 June 1945. Ibid., Minutes of Annual Meetings.
32 PANB, RS 136, File Q4, F.J. Elliott to J.A. Doucet, 21 June 1944.
33 CSLT Membership Files, M.D. to Ileen Kemp, 8 May 1952.
34 CSLT, Minutes of the Executive, 17 October 1943.
35 Harvey, "Nursing Service in the Small Hospital," 501. See also McPherson, "The Country Is a Stern Nurse," 175–206.
36 Susan Porter Benson, in her work on department-store sales clerks has written that:

> Workers and managers carry in their heads a cultural map of the work world which balkanizes it into enclaves according to sex, age, ethnicity, race, and class. But this map cannot predict what any individual will do for a living. Virtually no woman job-seeker between 1890 and 1940 had a statistician's grasp of the opportunities nor, even if she had, would she have based her decision on quantifiable variables alone. Her personal needs and inclinations guided her along with gossip about good jobs, good bosses, good workplaces, and available openings that circulated in family, neighborhood, and peer networks. (Benson, *Counter Cultures*, 181–2).

In Canada, Cecelia Reynolds has explored how gender, ethnicity, and class affected not only people's decisions to enter teaching but also their place within the educational hierarchy for the period 1930 to 1980. Reynolds, "Naming the Experience."

37 Laboratory work was not the only such niche. For example, Toronto's Hospital for Sick Children appointed a dietitian in 1908 and hospitals were the most important employers of dietitians. See Heap, "From the Science of Housekeeping," 157–8. Rima Apple has most explicitly explored the gendering of nutrition science. She argues that in the United States between 1840 and 1940, nutrition permitted some women a career in science when other avenues were closed to them. See "Science Gendered." American historian Margaret Rossiter explores the growth of home economics in her book *Women Scientists in America*, thereby firmly situating this work within the realm of science. In contrast, Marianne Gosztonyi Ainley's book *Despite the Odds* – which remains the only collection on women scientists in Canada – contains no articles that examine work in health care. In other words, the science of health care work was written out of this collection. In Canada, Kathryn McPherson has made the most cogent argument for a reconsideration of this position. See McPherson, *Bedside Matters*, 76.
38 Interview with Rose Phillips, conducted by Peter L. Twohig, 22 April 1996.
39 Ibid.
40 Interview with Edna Williams, conducted by Peter L. Twohig, 23 April 1996.
41 Ibid. See also Reverby, "Neither for the Drawing Room nor the Kitchen."
42 Interview with Ellen Robinson, conducted by Peter L. Twohig, 29 April 1996.
43 Ibid.
44 Botany enjoys a special place in the history of women's engagement with science. Rossiter, *Women Scientists in America*, 61, 233; Merchant, "Isis' Consciousness Raised," 398–409; Ainley, "Last in the Field?," 25–62; Gillett, "Carrie Derrick," 74–87.
45 CSLT, Membership Files.
46 NSARM, RG25, Series B, Vol. 3, BOC, 12 February 1931.
47 See JHA 1926, Nova Scotia, *Public Accounts*.
48 VGHL, W.W. Kenney to A.G. Nicholls, 3 December 1924, and W.W. Kenney to Mrs J. Nelson Gowanloch, 3 December 1924.
49 Gowanloch graduated from Rush Medical College in New York in August 1929 and interned at the Metropolitan and Bellevue Hospitals. She enrolled in Rush under her maiden name of Ross. See DUAPO, Staff Files, "Gowanloch, L.R."
50 Roper, "Two Scandals in Academe," 127–45.
51 Ibid.
52 Ibid.
53 Interview with Edna Williams, conducted by Peter L. Twohig, 23 April 1996.

54 CSLT, Membership Files, and interview with Laura Piers, conducted by Peter L. Twohig, 26 October 1996.
55 CSLT, Membership Files.
56 Ibid.
57 Ibid.
58 The best context for the war years is Pierson, *"They're Still Women After All."*
59 Interview with Edna Williams, conducted by Peter L. Twohig, 23 April 1996.
60 CSLT, Minutes of the Executive, 14 December 1940, and *Canadian Journal of Medical Technology* 2 (1940). The Royal Canadian Air Force ran an advertisement for twelve workers in this edition. Enlistees were promised "rapid advancement to corporal and sergeant." See also Blackwell, "A Military Hospital Laboratory," 16–19.
61 CSLT, Minutes of the Executive, 4 May 1940.
62 Siekawitch, Letter to the Editor, 146.
63 Twohig, "Culturing Professionalism," 61–3.
64 CSLT, Minutes of AGM, 13 June 1942 and 29 May 1943.
65 G.H. Agnew and W.J. Deadman to the Directors of Medical Service for the Navy (Surgeon Captain A. McCalum), Army (Brigadier G.B. Chisholm), and Air Force (Air Commodore R.W. Ryan), reproduced in CSLT, Minutes of the Executive, 17 October 1942.
66 CSLT, Minutes of AGM, 20 May 1944. For one perspective on laboratory services during war, see Bowman, "Laboratory Work in the Field," 34–5.
67 NAC, MG28-I343, CMA, Minutes of Executive Committee, Ottawa, 30–31 October 1941.
68 CSLT, Minutes of AGM, 11 December 1937.
69 CSLT, Minutes of the Executive, 10 May 1941. The applicants were from the military hospitals at Camp Borden and Kingston, Ontario, Calgary, and Esquimault.
70 Ibid., 31 January 1942.
71 Ibid., 11 March 1944.
72 G.H. Agnew and W.J. Deadman to the Directors of Medical Service for the Navy (Surgeon Captain A. McCalum), Army (Brigadier G.B. Chisholm), and Air Force (Air Commodore R.W. Ryan), reproduced in CSLT, Minutes of the Executive, 17 October 1942.
73 CSLT, Minutes of the Executive, 19 October 1940 and 14 December 1949.
74 CSLT, Correspondence Files, Harvey Agnew to Denys Lock, 31 October 1940. See also Lock to Agnew, 22 October 1940.
75 CSLT, Minutes of the Executive, 25 October 1945 and 6 April 1946.
76 Ibid., Detailed Agenda, 7 December 1952.
77 CSLT, Minutes of AGM, 1 June 1946 and 31 May 1947. The Grenfell Association, founded by Wilfred Grenfell in 1895, provided medical services in

northern Newfoundland and Labrador. See Rompkey, *Grenfell of Labrador*, particularly part four. For the broader context, see Neary, *Newfoundland in the North Atlantic World 1929–1949*, 52–4.

78 CSLT, Minutes of the Executive, 27 May 1951. In that year, there were fifty-seven approved schools across the country training laboratory workers, with an annual enrolment of about 175. The CSLT estimated that an additional two thousand technicians would be required by 1960, suggesting a shortfall of about eight hundred.

79 Ibid.

80 Interview with Rose Phillips, conducted by Peter L. Twohig, 22 April 1996.

81 Interview with Edna Williams, conducted by Peter L. Twohig, 23 April 1996.

82 Joan Sangster (*Earning Respect*, 78) has recently illustrated the impact of marriage on the working life of women. She wrote:

> The knowledge that a marriage bar existed – at least until the Second World War and in some factories until the 1950s – also put closure on the women's mental images of job change. This marriage bar was so pervasive in the local industries that some women believed "that was the law then." Moreover, the hostility sometimes directed by co-workers towards pregnant or married women (who in one factory were referred to as "money grubbers") did not encourage resistance to this barrier.

83 CSLT, Registry Files.

84 Cockburn, *Machinery of Dominance*, 136.

85 Bradbury, *Working Families*, 52.

86 CSLT, Minutes of AGM, 19 May 1952.

87 CSLT, Minutes of the Executive, 27 May 1951; my emphasis. The appeal for men also occurred in the early years of the twentieth century in Ontario's public schools. There school boards were encouraged to pay higher salaries to men to retain them, thereby ensuring the "quality of education" for older boys. See Gelman, "The 'Feminization' of the High Schools?," 119–48. Within health care, physiotherapy underwent a similar debate in the 1950s and 1960s, suggesting that recruiting men would enhance the professional status of physiotherapy. See Heap, "Physiotherapy's Quest," 73.

88 CSLT, Minutes of AGM, 19 May 1952.

89 Etzioni, *The Semi-Professions*.

90 CSLT, Minutes of the Executive, 7 December 1952.

91 "Report of the Committee on Laboratory Technicians," in NAC, MG28-1343, CMA, Minutes of Executive Committee, 3–4 March 1944.

92 CSLT, Minutes of the Executive, Meeting Detailed Agenda, "Report of the Committee on Public Relations."

93 *Canadian Nurse* 41 (September 1945): 696.

94 CSLT, Minutes of the Executive, 20–21 November 1954.
95 Canada, Department of Labour, *Medical Laboratory Technologist*. This publication was part of the "Canadian Occupations" series produced by the federal labour department.
96 CSLT, Minutes of AGM, 20 June 1956.
97 Ibid.
98 In 1952, for example, over 2,700 career sheets were mailed to every province. See CSLT, Minutes of the Executive, Detailed Agenda, 19 October 1952 and Minutes of the Executive, 19 October 1952. At the annual general meeting for 1953, it was reported that the sheets had been mailed to every high school in Canada. Ibid., Minutes of AGM, 18 May 1953.
99 CSLT, Correspondence Files, Ileen Kemp to Mrs D.B.M., 23 April 1958. Kemp said the national office tried to "distribute the career sheets fairly liberally and to reserve the pictorial pamphlets for the hands of the more interested students since it is a fairly costly item and we don't have the funds to pass it out as we can the career sheet." The CSLT prepared its own leaders for before and after the strip to "Canadianize" the strip and replaced the American recruitment film in 1965. In that year, the society produced a twenty-minute, 16mm colour recruitment film complete with narrated sound and background music, at a cost of $3,600. Shearer, "Canadian Society of Laboratory Technologists," 37; Canada, Department of Labour, *Medical Laboratory Technologist*.
100 Shearer, "Canadian Society of Laboratory Technologists," 15.
101 CSLT, Committee on Education, "If You Like Science."
102 CSLT, "Medical Technology: A Career with a Future."
103 Ontario Hospital Association, "Hospital Careers."
104 CSLT Committee on Education, "If You Like Science."
105 Canada, Department of Labour, *Medical Laboratory Technologist*.
106 Ibid.
107 As suggested in the previous chapters, attrition was high among laboratory workers. Those emerging from training programs were barely sufficient to replace those who left the workforce, either through emigration or marriage. At the same time, laboratories were being constructed or enlarged across the country and vacancies were common.
108 Canada, Department of Labour, *Medical Laboratory Technologist*.
109 Gerard and Kemp, "Medical Laboratory Technologist." This was a Guidance Centre Occupational Information Monograph, developed under the auspices of the Ontario College of Education, first published in 1954 and republished in 1958.
110 For the implications of this definition for women, see Cott, *The Grounding of Modern Feminism*, 216–17.
111 Cockburn, *Brothers*, 112–14.

112 "Menial" work is a pejorative term and I do not wish to use it uncritically. As Tim Diamond has argued in his analysis of contemporary nursing-home care, classifying some work as "menial" obscures the highly skilled and complex nature of caring for the elderly or the sick. Diamond, *Making Gray Gold*.
113 Coburn, et al., "Medical Dominance in Canada," 414.
114 Adams, *A Dentist and a Gentleman*, 110.
115 Ibid., 173.
116 Bruno Latour has recently written that "one cannot get much mileage" out of the term, because of the lack of specificity. "All kinds of people," he wrote, "are being professionalised during the nineteenth century, from bank-tellers to doctors." Similarly, the infrastructure that is developed – the buildings and organizations – is often portrayed as the "institutionalization" of these professions. Latour concludes that "every single discipline is doing the same in the nineteenth century: getting a profession, institutions and buildings. So it is a nineteenth-century feature; it is not specific." Latour, "The Costly Ghastly Kitchen," 295–6.
117 Geison, "Introduction," 3–4.
118 Scott, "The Profession That Vanished," 12–28. For the need to understand national contexts, see McClelland, "Escape from Freedom?," 97–113.
119 Torstendahl, "Introduction," 1–10; and Brumberg and Tomes, "Women in the Professions," 275–96.
120 Barker, "Women Physicians," 229–55.
121 Prentice and Theobald, *Women Who Taught*.
122 See, for example, Sandelowski, *Devices and Desires*; McPherson, *Bedside Matters*; Reverby, *Ordered to Care*; and Melosh, "The Physician's Hand."
123 Morantz-Sanchez, *Sympathy and Science*.
124 Heap, "Physiotherapy's Quest"; Heap, "Training Women"; and Heap and Stuart, "Nurses and Physiotherapists."
125 Twohig, "Once a Therapist, Always a Therapist."
126 Charles and Fahmy-Eid, "La Diététique et la Physiothérapie."
127 Harris, *Beyond Her Sphere*; Glazer and Slater, *Unequal Colleagues*; Fahmy-Eid, *Femmes, santé et professions*; Kinnear, *In Subordination*.
128 Smyth, et al., *Challenging Professions*.
129 Shearer, ed., "Canadian Society of Laboratory Technologists," 1. The British organization was renamed the Institute of Medical Laboratory Technology in 1942. The American Society was preceded by a Registry of Medical Technologists, which began in 1928 under the auspices of the American Society of Clinical Pathologists.
130 Cott, *The Grounding of Modern Feminism*, 230–1. The relationship between professionalism and feminism is explored in chapter 7 of Cott's work.

131 Kinnear, "Disappointment in Discourse," 282–3.
132 Cott, *The Grounding of Modern Feminism*, 231.
133 Kealey, "The Structure of Canadian Working Class History," 28. For additional Canadian context, see Lowe, *Women in the Administrative Revolution*, and Creese, *Contracting Masculinity*. Two outstanding American case studies of Pittsburgh and Chicago, respectively, are DeVault, *Sons and Daughters of Labor*, and Fine, *The Souls of the Skyscraper*.
134 Parr, *The Gender of Breadwinners*, 183.
135 Coburn, "Professionalization and Proletarianization," 140.
136 Brumberg and Tomes, "Women in the Professions," 276–7.
137 Kinnear, *In Subordination*, 152.
138 Ibid., 152, 162.
139 McPherson, *Bedside Matters*, 17.
140 Prognosticating about the future of laboratory diagnostic services, A.L. MacNabb classified workers as follows: the part-time worker in the small institution; the full-time assistant working in bacteriological or chemical testing; the senior worker who prepared tissue samples and undertook hematological tests and some bacteriological work; and the laboratory worker who graduated from "an arts faculty in which a course of instruction has been taken in biochemistry, haematology, and other related subjects, in connection with diagnostic procedures." MacNabb, "Possible Trends," 426.
141 The CMA Committee on Approval certified the Pathological Institute to provide classroom and practical training in laboratory technology in 1941. For information on the combined position at the Tuberculosis Hospital, see "News Notes," *Canadian Nurse* 45 (December 1949): 943.
142 Gerard and Kemp, *Medical Laboratory Technologist*.
143 CSLT, Minutes of AGM, 26 June 1950 and 7 June 1954. The pledge was modelled on an American document and had been in development for two years. See CSLT, Minutes of the Executive, 18 May 1952. The CSLT pledge declared:

> As I am about to commence my career as a medical technologist, and to assume its special responsibilities, I solemnly promise, before God:
>
> To carry out the duties assigned to me faithfully, and to the best of my ability.
>
> To remember that a patient's life may depend upon the accuracy and reliability of my work.
>
> To respect human life, and to be sympathetic towards all patients.
>
> To hold inviolate the confidence placed in me by both patient and doctor.
>
> To work harmoniously with my fellow technologists, and others who care for the sick.
>
> To uphold the ethics and dignity of my profession.
>
> To maintain an open mind for new ideas, and new truth.

144 CSLT, Minutes of AGM, 11 December 1937. See also McGhie, "The Laboratory," 36.
145 Gerard and Kemp, *Medical Laboratory Technologist*.
146 Ibid. The emphasis was in the original.
147 Hall, "Do's and Don'ts for Technicians," 51.
148 Miller, "The Characteristics and the Training of the Technologist," 43.
149 For a discussion of this issue, see Tillotson, "We May All Soon Be 'First Class Men,'" 97–125.
150 McPherson, *Bedside Matters*, 36–9, and ch 5, passim.
151 Korinek, *Roughing It in the Suburbs*, 297–9 and 338–40, and Kinsman, *The Regulation of Desire*.
152 Claussen, "Team Work," 145–8.
153 Graff, "Choosing a Technician," 68–70.
154 Miller, "The Characteristics and the Training of the Technologist," 42.
155 Ibid.
156 Hall, "Do's and Don'ts for Technicians," 52.
157 Miller, "The Characteristics and Training of the Technologist," 42.
158 Ibid., 43.
159 O'Donnell, "O Pity the Poor Student," 41–2.
160 Interview with Edna Williams, conducted by Peter L. Twohig, 23 April 1996.
161 CSLT, Membership Files, 1938.
162 Ibid., 1938.
163 Ibid., 1942.
164 Ibid., 1940.
165 Gleason, "Interim Report," 28.
166 CSLT, Minutes of the Executive, 14 October and 30 October 1944.
167 Ibid., 6 November 1943.
168 CSLT, Minutes of AGM, 20 May 1949.
169 CSLT, Minutes of the Executive, 17 October 1943.
170 McPherson, *Bedside Matters*, 230–4.
171 Sykes and Sethi, "The Labour Movement in Health Care: Canada," 42–53. See also White, *Hospital Strike*.
172 CSLT, Minutes of the Executive, 10 April 1949.
173 Ibid., 20 March 1954.
174 CSLT, Minutes of AGM, 6 June 1955.
175 Ibid.
176 Interview with Edna Williams, conducted by Peter L. Twohig, 23 April 1996.
177 Ibid.
178 Interview with Rose Phillips, conducted by Peter L. Twohig, 22 April 1996.
179 CSLT, Membership Files, 1950.
180 Davies, "The Sociology of Professions," 663.

CONCLUSION

1 REM, "King of Birds," Document (1986).
2 Deutsch, "Menace in the Medical Labs," 32–3, 89–90.
3 Tracey Adams cautions that "by focusing on only one profession of a relatively small size, one risks losing sight of the interdependence and interaction among professions." Adams, *A Dentist and a Gentleman*, 5.
4 For some recent examples that successfully transcend intellectual boundaries see Kelm, *Colonizing Bodies*; Lux, *Medicine That Walks*; and Anderson, *Vancouver's Chinatown*, especially chapters 3, 4, and 6.
5 Prentice, "Three Women in Physics," 120.
6 Larkin, *Occupational Monopoly and Modern Medicine*, vi.
7 This has been argued for a somewhat later period and in an office setting by Gillian Creese in *Contracting Masculinity*, 94.
8 See, for example, Cockburn, *Brothers*; Gaskell, "Conceptions of Skill," 11–25; Cockburn, "The Gendering of Jobs," 29–42; Parr, "Disaggregating the Sexual Division of Labour," 511–33; Parr, *The Gender of Breadwinners*. The best regional study of skill and gender is Tillotson, "The Operators along the Coast," 72–88.
9 I am here echoing Gregory Kealey's argument that Canadian workers have been central to Canadian historical development and that Canadian history cannot be understood without their inclusion. Kealey, "The Structure of Canadian Working-Class History," 23.
10 Apologies to Daniel Samson for paraphrasing the eloquent opening to his edited collection. See his "Introduction: Situating the Rural in Atlantic Canada," 1.
11 Larkin, *Occupational Monopoly and Modern Medicine*, 14.
12 Freidson, *Professional Powers*, 211.
13 Duffy, *The Sanitarians*, 195–6.
14 Irwin, "A Year of Challenge," 133.
15 CSLT, "On the Shoulders of Giants: A.R.," 105.
16 Ibid., "On the Shoulders of Giants: Norman Senn," 137–8. The CSLT reran a profile of Senn that it had previously published when Senn was completing his term as president of the society. Senn's comments are contained in the update to that original article.
17 In an unusual twist, Mallon began work during her second isolation. She started in the hospital in 1918, first as a domestic worker; then in 1922 she was termed a nurse and later became a "hospital helper." Beginning in 1925, she began to work in the hospital laboratory. Leavitt, *Typhoid Mary*, 193.
18 "1998 Planning Guide," 230–5.
19 Ibid., 230.
20 The slogan was for National Medical Laboratory Week, April 13–19, 1986.

21 For a local perspective, see "President's Message," 4–5.
22 Nova Scotia House of Assembly, *Debates and Proceedings*, April 27, 1995. The Institute of Technology's program also serviced students from Prince Edward Island. See also Nova Scotia Government Employees' Union *Newsletter* 95 (Summer 1995): 6.
23 Pierce, "What's In a Name?," 23.
24 Adams, *A Dentist and a Gentleman*, 171–2.
25 See, for example, Tillotson, "We May All Soon Be 'First Class Men.'"
26 For two excellent examples of how historians have paid attention to the role of place in women's health care work, and its interaction with other factors such as education, gender, ethnicity, and class, see Charlotte Borst, "The Training and Practice of Midwives," and Smith, "White Nurses, Black Midwives."
27 Parr, *The Gender of Breadwinners*, 231, 234.
28 Irwin, "A Year of challenge," 133.

Bibliography

MANUSCRIPT COLLECTION

CANADIAN SOCIETY OF LABORATORY TECHNOLOGISTS (CSLT)
Canadian Society of Laboratory Technologists National Office, Hamilton, Ontario

Correspondence Files
Membership Files
Minutes of Annual General Meetings (AGMs)
Minutes of the Executive
CSLT Registry

DALHOUSIE UNIVERSITY ARCHIVES (DUA)
Dalhousie University, Halifax, Nova Scotia

President's Office Staff Files (DUAPO)
President's Office Correspondence
Dalhousie University Student Registers

NATIONAL ARCHIVES OF CANADA (NAC)
Ottawa, Ontario

MG28-175. Canadian Lung Association fonds
MG28-175. Canadian Tuberculosis Association (CTA)
MG28-1343, microfilm reels 7486, 7487. Canadian Medical Association fonds
  - Minutes of Annual Meetings
  - Minutes of General Council

NOVA SCOTIA ARCHIVES AND RECORDS MANAGEMENT (NSARM)
Halifax, Nova Scotia

Barbara Keddy fonds, Series 018. Social History of Nursing in Nova Scotia in the 1930s
MG20, Vol. 197. Minutes of the Massachusettes-Halifax Health Commission (MHHC)
RG25, Series B, Vol. 2. Minutes of the Medical Board (MMB)
RG25, Series B, Vol. 3. Minutes of the Victoria General Hospital Board of Commissioners (BOC) 1910–1943
RG25, Series C, Vol. 1. Records of the Department of Public Health

PROVINCIAL ARCHIVES OF NEW BRUNSWICK (PANB)
Fredericton, New Brunswick

RS136. Records of the Deputy Minister of Health

QUEEN ELIZABETH II HEALTH SCIENCES CENTRE
RG25, Series B. Victoria General Hospital Letterbook (VGHL)

SECONDARY SOURCES

Adams, Tracey L. *A Dentist and a Gentleman: Gender and the Rise of Dentistry in Ontario*. Toronto: University of Toronto Press, 2000.
Agnew, G. Harvey. *Canadian Hospitals 1920 to 1970: A Dramatic Half Century*. Toronto: University of Toronto Press, 1974.
Ainley, Marianne Gosztonyi. "Last in the Field? Canadian Women Natural Scientists, 1815–1965." In *Despite the Odds: Essays on Canadian Women and Science*, edited by Marianne Gosztanyi Ainley, 25–62. Montreal: Vehicule Press, 1990.
Ainley, Marianne Gosztonyi, ed. *Despite the Odds: Essays on Canadian Women and Science*. Montreal: Vehicule Press, 1990.
Amirault, Marjorie Adams. "The Historical Evolution of Nursing Education in a Small Diploma School: 1913–1958." Master's of Nursing thesis, Dalhousie University, Halifax, Nova Scotia, 1991.
Anderson, Kay J. *Vancouver's Chinatown: Racial Discourse in Canada, 1875–1980*. Montreal: McGill-Queen's University Press, 1991.
Andrews, Margaret W. "The Best Advertisement a City Can Have: Public Health Services in Vancouver, 1886–1888," *Urban History Review* 12, no. 3 (February 1984): 19–27.
Annual Report of the Directors of the Chipman Memorial Hospital.
Annual Report of the Hotel Dieu Hospital, Chatham, New Brunswick.
Annual Report of the Miramichi Hospital.
Apple, Rima. "Science Gendered: Nutrition in the United States, 1840–1940." in Harmke Kamminga and Andrew Cunningham, eds., *The Science and Culture of Nutrition, 1840–1940*, edited by Haranke Kaminga and Andrew Cunninghan, 129–54. Amsterdam: Rodopi, 1995.

Archangelo, V., M. Fitzgerald, D. Carroll, and J.D. Plumb. "Collaborative Care between Nurse Practitioners and Primary Care Physicians." *Primary Care* 23, no. 1 (1996): 103–13.
Artibise, Alan F.J. *Winnipeg: A Social History of Urban Growth 1874–1914*. Montreal: McGill-Queen's University Press, 1975.
Axelrod, Paul. *Making a Middle Class: Student Life in English Canada During the Thirties*. Montreal: McGill-Queen's University Press, 1990.
Baker, Elizabeth F. *Technology and Woman's Work*. New York: Columbia University Press, 1964.
Baldwin, Douglas O. *She Answered Every Call: The Life of Public Health Nurse Mona Gordon Wilson (1894–1981)*. Charlottetown: Indigo Press, 1997.
– "Volunteers in Action: The Establishment of Government Health Care on Prince Edward Island, 1900–1931." *Acadiensis* 19 (Spring 1990): 121–47.
– "The Campaign against Odors: Sanitarians and the Genesis of Public Health in Charlottetown, Prince Edward Island (1855–1900)." *Scientia canadiensis* 10 (1986): 72–82.
Barley, Stephen R., and Julian E. Orr. "The Neglected Workforce." In *Between Craft and Science: Technical Work in US Settings*, edited by Stephen R. Barley and Julian E. Orr, 1–19. Ithaca: Cornell University Press, 1997.
Barker, Kristin. "Women Physicians and the Gendered System of Professions: An Analysis of the Shepppard-Towner Act of 1921." *Work and Occupations* 25, no. 2 (1998): 229–55.
Bator, Paul Adolphus. "'Saving Lives on the Wholesale Plan'": Public Health Reform in the City of Toronto, 1900 to 1930. PhD thesis. University of Toronto, 1979.
Beattie, Betsy. *Obligation and Opportunity: Single Maritime Women in Boston, 1870–1930*. Montreal and Kingston: McGill-Queen's University Press, 2000.
Beck, J. Murray. *Politics of Nova Scotia, Vol. 2: 1896–1988*. Tantallon: Four East Publications, 1985.
Begun, James W., and Ronald C. Lippincott. "The Origins and Resolution of Interoccupational Conflict." *Work and Occupations* 14, no. 3 (August 1987): 368–86.
Benison, Saul. "Poliomyelitis and the Rockefeller Institute: Social Effects and Institutional Response." *Journal of the History of Medicine and Allied Sciences* 29, no. 1 (January 1974): 74–92.
Benson, Susan Porter. *Counter Cultures: Saleswomen, Managers, and Customers in American Department Stores 1890–1940*. Urbana: University of Illinois Press, 1986.
Bickerton, James. "Reforming Health Care Governance: The Case of Nova Scotia." *Journal of Canadian Studies* 34, no. 2 (Summer 1999): 159–90.
Biggs Simon. "User Voice, Interprofessionalism and Postmodernity." *Journal of Interprofessional Care* 11, no. 2 (August 1997): 195–203.

Blackwell, Andrew. "A Military Hospital Laboratory." *Canadian Journal of Medical Technology* 6, no. 1 (1944): 16-19.
Bliss, Michael. *Banting: A Biography.* Toronto: McClelland and Stewart, 1984.
- *The Discovery of Insulin.* Toronto: McClelland and Stewart, 1982.
Bonner, Thomas Neville. *Medicine in Chicago 1850-1950: A Chapter in the Social and Scientific Development of a City.* Madison: The American History Research Center, 1957.
Borst, Charlotte. "The Training and Practice of Midwives: A Wisconsin Study." In *Women and Health in America*, edited by Judith Walzer Leavitt, 425-43. Madison: University of Wisconsin Press, 1999.
Bowman, F.W. "Laboratory Work in the Field." *Canadian Journal of Medical Technology* 2 (December 1939): 34-5.
Buckley, Suzann, and Janice Dickin McGinnis. "Venereal Disease and Public Health Reform in Canada." *Canadian Historical Review* 63 (1982): 337-54.
Bradbury, Bettina. *Working Families: Age, Gender and Daily Survival in Industrializing Montreal.* Toronto: McClelland and Stewart, 1993.
Brandt, Allan M., and David C. Sloane. "Of Beds and Benches: Building the Modern American Hospital." In *The Architecture of Science*, edited by Peter Galison and Emily Thompson, 281-305. Cambridge: MIT Press, 1999.
Brandt, Allan M. *No Magic Bullet: A Social History of Venereal Disease in the United States Since 1880.* New York: Oxford University Press, 1985.
Brown, Carol A. "The Division of Laborers: Allied Health Professions." In *Organization of Health Workers and Labor Conflict*, edited by Samuel Wolfe, 115-26. Farmingdale: Baywood Publishing, 1978.
Brouwer, Ruth Crompton. "Beyond 'Women's Work for Women': Dr Florence Murray and the Practice and Teaching of Western Medicine in Korea, 1921-1942." In *Challenging Professions: Historical and Contemporary Perspectives on Women's Professional Work*, edited by Elizabeth Smyth, Sandra Acker, Paula Bourne, and Alison Prentice, 65-95.
Bruce, Robert V. *The Launching of Modern American Science 1846-1876.* New York: Alfred A. Knopf, 1987.
Brumberg, Joan Jacob, and Nancy Tomes, "Women in the Professions: A Research Agenda for American Historians." *Reviews in American History* 10 (1982): 275-96.
Brush, B.L., and E.A. Capezuti. "Professional Autonomy: Essential for Nurse Practitioner Survival in the 21st Century." *Journal of the American Academy of Nurse Practitioners* 9, no. 6 (1997): 265-70.
Buchanan, L. "The Acute Nurse Practitioner in Collaborative Practice." *Journal of the American Academy of Nursing Practice* 8, no. 1 (1996): 13-20.
Burtch, Brian. *Trials of Labour: The Re-emergence of Midwifery.* Kingston and Montreal: McGill-Queen's University Press, 1994.
Buxton, William J. "Private Wealth and Public Health: Rockefeller Philanthropy and the Massachusetts-Halifax Relief Committee/Health Commission." In

Ground Zero: A Reassessment of the 1917 Explosion in Halifax Harbour, edited by Alan Ruffman and Colin D. Howell, 183–93. Halifax: Nimbus and Gorsebrook Research Institute, 1994.

Callbeck, J.A. Claudette. *A History of the PEI Hospital School of Nursing, 1891–1971*. Charlottetown: PEI Hospital, 1974.

Cameron, Ian A. "Camp Hill and the Smallpox Outbreak of 1938." *Nova Scotia Medical Bulletin* (June 1988): 100–3.

– "The Quarantine Station on Lawlor's Island 1866–1938." *Nova Scotia Medical Bulletin* (June-August 1983): 83–7.

Cameron, James. *"And Martha Served": History of the Sisters of St Martha, Antigonish, Nova Scotia*. Halifax: Nimbus, 2000.

– *For the People: A History of St Francis Xavier University*. Montreal: McGill-Queen's University Press, 1996.

Campbell, P.S., and H.L. Scammell. "The Development of Public Health in Nova Scotia." *Canadian Journal of Public Health* 30 (May 1939): 226–38.

Canada. Commission of Inquiry on the Blood System in Canada. *Final Report*. Ottawa: The Commission, 1997.

– Department of Labour. *Medical Laboratory Technologist*. Ottawa: Queen's Printer, 1957.

Canadian Society of Laboratory Technologists. "On the Shoulders of Giants: A.R. (Archie) Shearer." *Canadian Journal of Medical Technology* 58 (1996): 105.

– "On the Shoulders of Giants: Norman Senn." *Canadian Journal of Medical Technology* 58 (1996): 137–8.

Canadian Society for Laboratory Technology. "Medical Technology: A Career with A Future." January 1959.

– "If You Like Science Why Not Be a ... Medical Laboratory Technologist?" CSLT Committee on Education, May 1952.

Cassel, Jay. "Public Health in Canada." *Clio Medica* 26 (1994): 276–312.

– *The Secret Plague: Venereal Disease in Canada, 1838–1939*. Toronto: University of Toronto Press, 1987.

Cavanagh, S.J., and M. Bamford. Substitution in Nursing Practice: Clinical, Management and Research Implications. *Journal of Nursing Management* 5, no. 6 (November 1997): 333–9.

Charles, Aline, and Nadia Fahmy-Eid. "La Diététique et la Physiothérapie Face au Problème des Frontières Interprofessionnelles (1950–1980)." *Revue d'Histoire de l'Amérique Française* 47 (1994): 377–408.

Claussen, Frieda H. "Team Work – A Time Saver." *Canadian Journal of Medical Technology* 7, no. 4 (1945): 145–8.

Coburn, David. "Professionalization and Proletarianization: Medicine, Nursing, and Chiropractic in Historical Perspective." *Labour/Le Travail* 34 (Fall 1994): 139–62.

Coburn, David, George M. Torrance, and Joseph M. Kaufert. "Medical Dominance in Canada in Historical Perspective: The Rise and Fall of Medicine?" *International Journal of Health Services* 13 (1983): 407–32.

Cockburn, Cynthia. *Machinery of Dominance: Women, Men, and Technical Know-How*. Boston: Northwestern University Press, 1988.

– "The Gendering of Jobs: Workplace Relations and the Reproduction of Sex Segregation." In *Gender Segregation at Work*, edited by Sylvia Walby, 29–42. Philadelphia: Open University Press, 1988.

– *Brothers: Male Dominance and Technological Change*. London: Pluto Press, 1983.

Cogswell, Kathleen E. *Western Kings Memorial Hospital: The First Sixty Years 1922–1982*. New Minas: R.S. Babcock, [1982?].

Coleman, William, and Frederic L. Holmes, eds. *The Investigative Enterprise: Experimental Physiology in Nineteenth-Century Medicine*. Berkeley: University of California Press, 1988.

Collard, Patrick. *The Development of Microbiology*. Cambridge: Cambridge University Press, 1976.

Connor, J.T.H. "Hospital History in Canada and the United States." *Canadian Bulletin of Medical History* 7 (1990): 93–104.

Cooke, H. "Boundary Work in the Nursing Curriculum: The case of Sociology." *Journal of Advanced Nursing* 18 (December 1993): 1,990–8.

Copp, Terry. *The Anatomy of Poverty: The Condition of the Working Class in Montreal, 1897–1929*. Toronto: McClelland and Stewart, 1974.

Cott, Nancy F. *The Grounding of Modern Feminism*. New Haven: Yale University Press, 1987.

Cowan, Ruth Schwartz. *A Social History of American Technology*. Toronto: Oxford University Press, 1997.

– *More Work for Mother: The Ironies of Household Technology from the Open Hearth to the Microwave*. New York: Basic Books, 1983.

Cox, George H., and John H. MacLeod. *Consumption: Its Cause, Prevention and Cure*. London: Eyre and Spottiswoode, for the Tri-County Anti-Tuberculosis League, 1911.

Cunningham, Andrew, and Percy Williams. "Introduction." In *The Laboratory Revolution in Medicine*, edited by Andrew Cunningham and Perry Williams, 1–13. Cambridge: Cambridge University Press, 1992.

Creese, Gillian. *Contracting Masculinity: Gender, Class and Race in a White-Collar Union, 1944–94*. Don Mills: Oxford University Press, 1999.

Curtis, Bruce. *The Politics of Population: State Formation, Statistics, and the Census of Canada, 1840–1875*. Toronto: University of Toronto Press, 2001.

Daniel, J.W. "The Milk Supply and Its Control." *Maritime Medical News* 18, no. 12 (December 1906): 461–8.

Daniels, Arlenne Kaplan. "Invisible Work." *Social Problems* 34 (1987): 403–15.

Davies, Celia. "The Sociology of Professions and the Profession of Gender." *Sociology* 30 (November 1996): 661–78.

Defries, R.D. *The Development of Public Health in Canada: A Review of the History and Organization of Public Health in the Provinces of Canada.* Toronto: Canadian Public Health Association, 1940.

Deutsch, Albert. "Menace in the Medical Labs." *Woman's Home Companion* 77 (April 1950): 32–3, 89–90.

Devault, Ileen. *Sons and Daughters of Labor: Class and Clerical Work in Turn-of-the-Century Pittsburgh.* Ithaca: Cornell University Press, 1990.

Diamond, Timothy. *Making Gray Gold: Narratives of Nursing Home Care.* Chicago: University of Chicago Press, 1992.

DiCenso, A. "The Neonatal Nurse Practitioner." *Current Opinion in Pediatrics* 10, no. 2 (April 1998): 151–5.

Duffy, John. *The Sanitarians: A History of American Public Health.* Urbana: University of Illinois Press, 1990.

Dunlop Allan C. "The Collection of Vital Statistics in Nova Scotia, 1864–1908." *Royal Nova Scotia Historical Society Journal* 5 (2002): 134–52.

Elliot, Frank J. "Open Letter to All Members from the President." *Canadian Journal of Medical Technology* 1 (October 1938): 6.

Emery, George. *Facts of Life: The Social Construction of Vital Statistics, Ontario 1869–1952.* Montreal and Kingston: McGill-Queen's University Press, 1993.

Etzioni, Amitai, ed. *The Semi-Professions and Their Organization.* New York: Free Press, 1969.

Fahmy-Eid, Nadia, Aline Charles, Johanne Collin, Johanne Daigle, Pauline Fahmy, Ruby Heap, and Lucie Piché. *Femmes, santé et professions: histoire des diététistes et des physiothérapeutes au Québec et en Ontario, 1930–1980.* Saint-Laurent: Fides, 1997.

Farrell, Marilyn, "The Walkerton Tragedy: A Technologist's Story." *Canadian Journal of Medical Laboratory Science* 63, no. 2 (2001): 64–5.

Fine, Lisa. *The Souls of the Sykscraper: Female Clerical Workers in Chicago, 1870–1930.* Philadelphia: Temple University Press, 1990.

Forbes, E.R. "The Ideas of Carol Bacchi and the Suffragists of Halifax: A Review Essay on *Liberation Deferred? The Ideas of the English-Canadian Suffragists, 1877–1918.*" *Atlantis* 10 (Spring 1985): 119–26.

– "Prohibition and the Social Gospel in Nova Scotia." *Acadiensis* 1 (Autumn 1971): 11–36.

Francis, Daniel. "The Development of the Lunatic Asylum in the Maritime Provinces." *Acadiensis* 6 (Spring 1977): 23–38.

Fraser, N.S. "The Microscope in Diagnosis." *Maritime Medical News* 14 (May 1902): 162.

Freidson, Eliot. *Professional Powers: A Study of the Institutionalization of Formal Knowledge.* Chicago: University of Chicago Press, 1988.

– *Professional Dominance: The Social Structure of Medical Care.* New York: Atherton Press, 1970.
Fye, W. Bruce. *The Development of American Physiology 1790–1855.* Baltimore: Johns Hopkins University Press, 1987.
Gagan, David, and Rosemary Gagan. *For Patients of Moderate Means: A Social History of the Voluntary Public General Hospital in Canada, 1890–1950.* Montreal and Kingston: McGill-Queen's University Press, 2002.
Gagnon, Eugene. "Notes on the Early History and Evolution of the Department of Health of Montreal." *Canadian Journal of Public Health* 29 (May 1938): 216–23.
Garrett, Laurie. *The Coming Plague: Newly Emerging Diseases in a World Out of Balance.* New York: Farrar, Staus and Giroux, 1994.
Gaskell, Jane. "Conceptions of Skill and the Work of Women: Some Historical and Political Issues," *Atlantis* 8 (1983): 11–25.
Geison, Gerald L. *The Private Science of Louis Pasteur.* Princeton: Princeton University Press, 1995.
– "Introduction." In *Professions and Professional Ideologies in America,* edited by Gerald L. Geison. Chapel Hill: The University of North Carolina Press, 1983.
Geison, Gerald, ed. *Physiology in the American Context 1850–1940.* Baltimore: The American Physiology Society, 1987.
Gelman, Susan. "The 'Feminization' of the High Schools? Women Secondary School Teachers in Toronto: 1871–1930." *Historical Studies in Education/Revue D'Histoire de L'Education* 2 (1990): 119–48.
Gerard, Agnes, and Ileen Kemp. "Medical Laboratory Technologist." Toronto: The Guidance Centre, Ontario College of Education, 1958.
Gerard, Sister Catherine. "We Look at Nursing Service." *Canadian Nurse* 44 (October 1948): 827.
Gilbert, Leah. "Dispensing Doctors and Prescribing Pharmacists: A South African Perspective." *Social Science and Medicine* 46, no. 1 (January 1998): 83–95.
Gill, Mary. *75 Years of Caring: A History of the Miramichi Hospital, Newcastle, NB 1915–1990.* Chatham: Gemini Printing, 1990.
Gillett, Margaret. "Carrie Derrick (1862–1941) and the Chair of Botany at McGill." In Ainley, ed., *Despite the Odds: Essays on Canadian Women and Science,* edited by Marianne Gusztonyi Ainley, 74–87.
Glazer, Penina Migdal, and Miriam Slater. *Unequal Colleagues: The Entrance of Women into the Professions, 1890–1940.* New Brunswick: Rutgers University Press, 1987.
Gleason, Margaret. "Interim Report on a Survey Being Conducted to Determine Why Technicians Have Such Dispositions." *Canadian Journal of Medical Technology* 8 (1946): 27–8.

Godfrey, W.G. *The Struggle to Serve: A History of the Moncton Hospital, 1895 to 1953*. Montreal: McGill-Queen's University Press, 2004.
Godfrey, W.G. "Private and Government Funding: The Case of the Moncton Hospital, 1898–1953." *Acadiensis* 31 (Autumn 2001): 3–34.
Gorz, André. "Technology, Technicians and Class Struggle." In *The Division of Labour: The Labour Process and Class Struggle in Modern Capitalism*, edited by André Gorz. Atlantic Highlands, NJ: Humanities Press, 1976.
Graff C., and R.M. Forcade. "Choosing a Technician." *Canadian Journal of Medical Technology* 8, no. 2 (1946): 68–70.
Growe, Sarah Jane. *Who Cares? The Crisis in Canadian Nursing*. Toronto: McClelland and Stewart, 1991.
Hall, Harvey. "Do's and Don'ts for Technicians." *Canadian Journal of Medical Technology* 1, no. 2 (1939): 51–2.
Halpern Sydney A. "Medicalization as Professional Process: Postwar Trends in Pediatrics." *Journal of Health and Social Behavior* 31, no. 1 (March 1990): 28–42.
– *American Pediatrics: The Social Dynamics of Professionalism, 1880–1980*. Berkeley: University of California Press, 1988.
Harris, Barbara J. *Beyond Her Sphere: Women and the Professions in American History*. Westport, CT: Greenwood Press, 1978.
Hartmann, Heidi. "Capitalism, Patriarchy and Job Segregation by Sex." In *Capitalist Patriarchy and the Case for Socialist Feminism*, edited by Zilla R. Eisenstein. New York: Monthly Review Press, 1979.
Harvey, Kathleen. "Nursing Service in the Small Hospital." *Canadian Nurse* 33 (October 1937): 500–2.
Hayter, Charles. "To the Relief of Malignant Diseases of the Poor." *Journal of the Royal Nova Scotia Historical Society* 1 (1998): 130–43.
– "New Brunswick's Remarkable Dr Roberts." *Canadian Medical Association Journal* 146, no. 9 (1992): 1,637–9.
Heaman, E.A. "Review of Christopher J. Rutty, *A Circle of Care*," *Canadian Bulletin of Medical History*, 16, 1 (1999): 163–8.
Heap, Ruby. "From the Science of Housekeeping to the Science of Nutrition: Pioneers in Canadian Nutrition and Dietetics at the University of Toronto's Faculty of Household Science, 1900–1950." In *Challenging Professions: Historical and Contemporary Perspectives on Women's Professional Work*, edited by Elizabeth Smyth et al., 141–70.
– "Physiotherapy's Quest for Professional Status in Ontario, 1950–80." *Canadian Bulletin of Medical History* 12 (1995): 69–99.
– "Training Women for a New 'Women's Profession': Physiotherapy Education at the University of Toronto, 1917–40." *History of Education Quarterly* 35, no. 2 (1995): 135–58.

Heap, Ruby, and Meryn Stuart. "Nurses and Physiotherapists: Issues in the Professionalization of Health Care Occupations During and After World War I." *Health and Canadian Society* 2, nos. 1–2 (1995): 179–93.

Holmes, Frederic Lawrence. *Claude Bernard and Animal Chemistry: The Emergence of a Scientist*. Cambridge: Harvard University Press, 1974.

Howell, Colin D. "Medical Science and Social Criticism: Alexander Peter Reid and the Ideological Origins of the Welfare State in Canada." In *Canadian Health Care and the State: A Century of Evolution*, edited by C. David Naylor, 16–37. Montreal: McGill-Queen's University Press, 1992.

– *A Century of Care: A History of the Victoria General Hospital in Halifax, 1887–1987*. Halifax: Victoria General Hospital, 1988.

Howell, Joel D. *Technology in the Hospital: Transforming Patient Care in the Early Twentieth Century*. Baltimore: Johns Hopkins University Press, 1995.

Hughes, Everett C. *The Sociological Eye*. 2 Vols. Chicago: Aldine-Atherton, 1971.

– *Men and Their Work*. Glencoe, IL: Free Press, 1958.

Irwin, Jeanne. "A Year of Challenge – A Year of Change." *Canadian Journal of Medical Technology* 58 (1996): 133.

Ikeda, Kano. "Survey of Training Schools for Laboratory Technicians." *American Journal of Clinical Pathology* 1 (1931): 467–76.

Jacyna, L.S. "The Laboratory and the Clinic: The Impact of Pathology on Surgical Diagnosis in the Glasgow Western Infirmary, 1875–1910." *Bulletin of the History of Medicine* 62 (1988): 384–406.

Katz, Michael. "The Emergence of Bureaucracy in Urban Education: The Boston Case, 1850–1855." *History of Education Quarterly* (Summer 1968): 155–87.

Kealey, Gregory S. "The Structure of Canadian Working-Class History." In W.J.C. Cherwinski and Gregory S. Kealey, eds., *Lectures in Canadian Labour and Working-Class History*, edited by W.J.C. Cherwinski and Gregory S. Kealey, 23–36. St John's: Canadian Committee on Labour History, 1985. Reprinted in Gregory S. Kealey. *Workers and Canadian History*, 329–44. Montreal: McGill-Queen's University Press, 1995.

Keefe, Jeffrey, and Denise Potosky. "Technical Dissonance: Conflicting Portraits of Technicians." In *Between Craft and Science: Technical Work in US Settings*, edited by Stephen R. Barley and Julian E. Orr, 53–81. Ithaca: Cornell University Press, 1997.

Keddy, Barbara A. "Private Duty Nursing Days of the 1920s and 1930s in Canada." *Canadian Woman Studies/Les Cahier de la Femme* 7 (1984): 99–102.

Kelm, Mary-Ellen. *Colonizing Bodies: Aboriginal Health and Healing in British Columbia, 1900–50*. Vancouver: UBC Press, 1998.

Kinnear, Mary. *In Subordination: Professional Women 1870–1970*. Montreal: McGill-Queen's University Press, 1995.

- "Disappointment in Discourse: Women University Professors at the University of Manitoba Before 1970." *Historical Studies in Education/Revue Histoire de L'Education* 4 (1992): 269–87.
Kinsman, Gary. *The Regulation of Desire: Sexuality in Canada.* Montreal: Black Rose Books, 1987.
Kitz, Janet F. *Shattered City: The Halifax Explosion and the Road to Recovery.* Halifax: Nimbus, 1989.
Kohler, Robert E. *From Medical Chemistry to Biochemistry: The Making of a Biomedical Discipline.* New York: Cambridge University Press, 1982.
Korinek, Valerie J. *Roughing It in the Suburbs: Reading Chatelaine Magazine in the Fifties and Sixties.* Toronto: University of Toronto Press, 2000.
Larkin, Gerald. *Occupational Monopoly and Modern Medicine.* London: Tavistock Publications, 1983.
Larson, Magali Sarfatti. *The Rise of Professionalism: A Sociological Analysis.* Berkeley: University of California Press, 1977.
Latour, Bruno. "The Costly Ghastly Kitchen." In *The Laboratory Revolution in Medicine,* edited by Andrew Cunningham and Percy Williams, 295–303. Cambridge: Cambridge University Press, 1992.
Latour, Bruno, and Steve Woolgar. *Laboratory Life: The Social Construction of Scientific Facts.* Beverly Hills: Sage, 1979.
Leathard, A. *Going Inter-Professional: Working Together for Health and Welfare.* London: Routledge, 1994.
Leavitt, Judith Walzer. *Typhoid Mary: Captive to the Public's Health.* Boston: Beacon Press, 1996.
Linkletter, Lindsay. "An Open Door: Politics and Science in the Career of Alexander Peter Reid, 1878–92." MA thesis. Dalhousie University, Halifax, Nova Scotia, 1996.
Lonsdale, Susan, Adrian Webb, and Thomas L. Briggs. *Teamwork in the Personal Social Services and Health Care: British and American Perspectives.* London: Croon Helm, 1980.
Lowe, Graham. *The Administrative Revolution: The Feminization of Clerical Work.* Toronto: University of Toronto Press, 1987.
Lown, Judy. "Not So Much a Factory, More a Form of Patriarchy: Gender and Class During Industrialisation." In *Gender, Class and Work,* edited by Eva Gamannikow. Aldershot: Grover, 1985.
Lux, Maureen K. *Medicine that Walks: Disease, Medicine, and Canadian Plains Native People, 1880–1940.* Toronto: University of Toronto Press, 2001.
MacLellan, Donald I. *A History of the Moncton Hospital: A Proud Past – A Healthy Future (1895–1995).* Halifax: Nimbus Publishing, 1998.
MacDonald, A.H. *Mount Hope Then and Now: A History of the Nova Scotia Hospital.* Dartmouth: The Nova Scotia Hospital, 1996.

MacDonald, Lewis R. "Golden Gleanings: Commemorating the Fiftieth Anniversary of St Joseph's Hospital 1902–1952 and Its School of Nursing 1905–1955." Glace Bay, NS: St Joseph's Hospital, n.d. [1955?].

MacDougall, Heather. *Activists and Advocates: Toronto's Health Department, 1883–1983*. Toronto: Dundurn, 1990.

– "'Health is Wealth': The Development of Public Health Activity in Toronto, 1834–1890." PhD thesis, University of Toronto, 1981.

Mack, Frank G. "The General Practitioner and Urological Problems." *Nova Scotia Medical Bulletin* 5 (March 1926): 6–10.

MacKenzie, D.J. "The Origin and Development of a Medical Laboratory Service in Halifax." *Nova Scotia Medical Bulletin* 43 (1964): 179–84.

– "Some Phases of Poliomyelitis, *Nova Scotia Medical Bulletin* 8 (September 1929): 415–17.

MacLean, Terrence D. *Asylum: A History of the Cape Breton Hospital 1906–1995*. Sydney: Cape Breton Mental Health Services Charitable Foundation, 1996.

MacLeod, Enid Johnson. *Petticoat Doctors: The First Forty Years of Women in Medicine at Dalhousie University*. Porter's Lake: Pottersfield Press, 1990.

MacNabb, A.L. "Possible Trends in the Public Health Laboratory Diagnostic Service." *Canadian Journal of Public Health* 38 (September 1947): 422–7.

Mandel, Ernest. *Late Capitalism*. Atlantic Highlands, NJ: Humanities Press, 1975.

Martin, Paul. "Federal Aid Towards the Training of Nurses." *Canadian Nurse* 47 (May 1951): 327–30.

Mazumdar, Pauline M.H. *Immunology 1930–1980: Essays on the History of Immunology*. Toronto: Wall and Thompson, 1989.

McClelland, Charles E. "Escape from Freedom? Reflections on German Professionalization, 1870–1933." In *The Formation of Professions: Knowledge, State and Strategy*, edited by Rolf Torstendahl and Michael Burrage, 97–113. London: Sage Publications, 1990.

McCuaig, Katherine. *The Weariness, The Fever, and the Fret: The Campaign Against Tuberculosis in Canada, 1900–1950*. Montreal and Kingston: McGill-Queen's University Press, 1999.

– "From Social Reform to Social Service: The Changing Role of Volunteers in the Anti-Tuberculosis Campaign, 1900–1930." *Canadian Historical Review* 61 (1980): 480–501.

McGee, Arlee. *The Victoria Public Hospital, Fredericton: 1888–1976*. Fredericton: Victoria Public Hospital Nurses' Alumnae, 1984.

McGhie, B.T. "The Laboratory in Relation to Public Health." *Canadian Journal of Medical Technology* 1, no. 2 (March 1939): 36–41.

McKay, Ian. "A Note on 'Region' in Writing the History of Atlantic Canada." *Acadiensis* 29 (Spring 2000): 89–101.

- "The Stillborn Triumph of Progressive Reform." In *The Atlantic Provinces in Confederation*, edited by E.R. Forbes and D.A. Muise, 192–229. Toronto and Fredericton: University of Toronto and Acadiensis Press, 1993.
McPherson, Kathryn. *Bedside Matters: The Transformation of Canadian Nursing, 1900–1990*. Toronto: Oxford University Press, 1996.
- "The Country is a Stern Nurse: Rural Women, Urban Hospitals and the Creation of a Western Canadian Workforce, 1920–1940," *Prairie Forum*, 20 (Fall 1995): 175–206.
- "Skilled Service and Women's Work: Canadian Nursing 1920–1939." PhD thesis. Simon Fraser University, Vancouver, British Columbia, 1989.
McPherson, Kathryn M. "Nurses and Nursing in Early Twentieth Century Halifax." MA thesis. Dalhousie University, Halifax, Nova Scotia, 1982.
McPherson, Kathryn, and Meryn Stuart. "Writing Nursing History in Canada: Issues and Approaches." *Canadian Bulletin of Medical History* 11 (1994): 3–22.
Melosh, Barbara. *"The Physicians Hand": Work Culture and Conflict in American Nursing*. Philadelphia: Temple University Press, 1982.
Merchant, Carolyn. "Isis' Consciousness Raised." *Isis* 73 (1982): 398–409.
Mesler, Mark A. "Boundary Encroachment and Task Delegation: Clinical Pharmacists on the Medical Team." *Sociology of Health and Illness* 13, 3 (September 1991): 310–31.
Miller, James. "The Characteristics and the Training of the Technologist." *Canadian Journal of Medical Technology* 1 (March 1939): 42–6.
Mitchinson, Wendy. *Giving Birth in Canada, 1900–1950*. Toronto: University of Toronto Press, 2002.
Montgomery, David. *The Fall of the House of Labor: The Workplace, the State and American Labor Activism, 1865–1925*. Cambridge: Cambridge University Press, 1987.
- *Workers' Control in America: Studies in the History of Work, Technology, and Labor Struggles*. Cambridge: Cambridge University Press, 1979.
Moran, James E. *Committed to the State Asylum: Insanity and Society in Nineteenth-Century Quebec and Ontario*. Montreal and Kingston: McGill-Queen's University Press, 2000.
Morantz-Sanchez, Regina Markell. *Sympathy and Science: Women Physicians in American Medicine*. New York: Oxford University Press, 1985.
Morrison, Pearl L. "The Nurses in Hospital Administration." *Canadian Nurse* 36 (October 1940): 672–4.
Morse, L.R. "Presidential Address." in *Nova Scotia Medical Bulletin* 1 (December 1922): 6.
Morton, Suzanne. *Ideal Surroundings: Domestic Life in a Working-Class Suburb in the 1920s*. Toronto: University of Toronto Press, 1995.
Morton, Suzanne. "The June Bride as the Working-Class Bride: Getting Married in a Halifax Working-Class Neighbourhood in the 1920s." In *Canadian*

Family History, edited by Bettina Bradbury, 360–79. Mississagua: Copp Clark Pitman, 1992.
Mukerji, Chandra. *A Fragile Power: Scientists and the State.* Princeton: Princeton University Press, 1989.
Murray, T.J. "The Visit of Abraham Flexner to Halifax Medical College." *Nova Scotia Medical Bulletin* 64 (June 1985): 34–41.
Naylor, C. David. *Private Practice, Public Payment: Canadian Medicine and the Politics of Health Insurance 1911–1966.* Montreal: McGill-Queen's University Press, 1986.
Neary, Peter. *Newfoundland in the North Atlantic World 1929–1949.* Montreal: McGill-Queen's University Press, 1988.
New Brunswick. *Annual Report of the Chief Medical Officer to the Minister of Health.* 1918–1945. Fredericton: Government of New Brunswick.
– *Annual Report of the Bureau of Laboratories.* 1918–1945. Fredericton: Government of New Brunswick.
– *Public Accounts.* 1919–20 to 1944–45. Fredericton: Government of New Brunswick.
Nicholson, Daniel. *Laboratory Medicine: A Guide for Students and Practitioners.* Philadelphia: Lea and Febiger, 1930.
Nova Scotia. *Department of Public Health Annual Report.* Halifax: Kings Printer, 1904–1945.
– "Report of the Provincial Board of Health." In *Journal of the House of Assembly.* Halifax: Kings Printer, 1893–1903.
O'Donnell, Mary W. "O Pity the Poor Student – Or Should We?" *Canadian Journal of Medical Technology* 4, no. 1 (1942): 38–42.
Ontario Hospital Association. "Hospital Careers: Opportunities Youth." Toronto: Ontario Hospital Association, n.d. [1958?].
Ovretveit, John. *Co-ordinating Community Care: Multidisciplinary Teams and Care Management.* Buckingham, UK: Open University Press, 1992.
Owens, Patricia, John Carrier, and John Horder. *Interprofessional Issues in Community and Primary Health Care.* London: Macmillan, 1995.
Overduin, Hendrick. *People and Ideas: Nursing at Western 1920–1970.* London: University of Western Ontario Faculty of Nursing, 1970.
Parr, Joy. *The Gender of Breadwinners: Women, Men, and Change in Two Industrial Towns 1880–1950.* Toronto: University of Toronto Press, 1990.
– "Disaggregating the Sexual Division of Labour: A Transatlantic Case Study." *Comparative Studies in Society and History* 30 (1988): 511–33.
Paul, John R. *A History of Poliomyelitis.* New Haven: Yale University Press, 1971.
Penney, Sheila M. "Inventing the Cure: Tuberculosis in 20th Century Nova Scotia." PhD thesis. Dalhousie University, Halifax, Nova Scotia, 1991.
– "'Marked for Slaughter': The Halifax Medical College and the Wrong Kind of Reform, 1868–1910." *Acadiensis* 19 (Fall 1989): 27–51.

- "Tuberculosis in Nova Scotia." MA thesis. Dalhousie University, Halifax, Nova Scotia, 1985.
Pharos (Dalhousie University Yearbook).
Picard, André. *The Gift of Death: Confronting Canada's Tainted Blood Tragedy*. Toronto: Harper Collins, 1995.
Pierce, Ruth. "What's In a Name?" *Nova Scotia Society of Medical Laboratory Technologists News and Views* (December 1996).
Pierson, Ruth Roach. *"They're Still Women After All": The Second World War and Canadian Womanhood*. Toronto: McClelland and Stewart, 1986.
Pothier, Evangeline R. *Mary Ann Watson and the Yarmouth Hospital*. [Yarmouth]: s.n., [1986].
Prentice, Alison. "Three Women in Physics." In *Challenging Professions: Historical and Contemporary Perspectives on Women's Professional Work*, edited by Elizabeth Smyth et al., 119–40.
Prentice, Alison, and Marjorie Theobald. *Women Who Taught: Perspectives on the History of Women and Teaching*. Toronto: University of Toronto Press, 1991.
Preston, Richard. *The Hot Zone*. New York: Random House, 1994.
Reaume, Geoffrey. *Remembrance of Patients Past: Patient Life at the Toronto Hospital for the Insane, 1870–1940*. Don Mills: Oxford University Press, 2000.
Reid, A.P. "Public Health." *Maritime Medical News* 13 (August 1901): 285–6.
Reid, John G. *Six Crucial Decades: Times of Change in the History of the Maritimes*. Halifax: Nimbus Publishing, 1987.
Reid, John G. "Health, Education, Economy: Philanthropic Foundations in the Atlantic Region in the 1920s and 1930s." *Acadiensis* 14 (Autumn 1984): 64–83.
Reiser, Stanley Joel. *Medicine and the Reign of Technology*. Cambridge: Cambridge University Press, 1978.
Reverby, Susan. "'Neither for the Drawing Room nor the Kitchen': Private Duty Nursing in Boston, 1873–1920." In *Sickness and Health in America: Readings in the History of Medicine and Public Health*, edited by Judith Walzer Leavitt and Ronald L. Numbers, 253–65. Madison: University of Wisconsin Press, 1997.
- "The Search for a Hospital Yardstick: Nursing and the Rationalization of Hospital Work." In *Health Care in America: Essays in Social History*, edited by Susan Reverby and David Rosner, 206–16. Philadelphia: Temple University Press, 1979.
Reverby, Susan M. *Ordered to Care: The Dilemma of American Nursing, 1850–1945*. Cambridge: Cambridge University Press, 1987.
Reynolds, Cecelia. "Naming the Experience: Women, Men and Their Changing Work Lives as Teachers and Principals." PHD thesis. University of Toronto, 1987.

Riska, Elianne, and Katarina Wegar. *Gender, Work and Medicine: Women and the Medical Division of Labour.* London: Sage, 1993.

Ritchey, Ferris J., and David G. Sommers. "Medical Rationalization and Professional Boundary Maintenance: Physicians and Clinical Pharmacists." *Research in the Sociology of Health Care* 10 (1993): 117-39.

Rogers, Naomi. *Dirt and Disease: Polio before FDR.* New Brunswick: Rutgers University Press, 1992.

Rompkey, Ronald. *Grenfell of Labrador: A Biography.* Toronto: University of Toronto Press, 1991.

Roper, Henry. "Two Scandals in Academe." *Collections of the Royal Nova Scotia Historical Society* 43 (1991): 127-45.

Rose, Sonya. *Limited Livelihoods: Gender and Class in Nineteenth-Century England.* Berkeley: University of California Press, 1992.

Rosen, George. *The Structure of American Medical Practice 1875-1941.* Philadelphia: University of Pennsylvania Press, 1983.

Rosenberg, Charles E. "Community and Communities: The Evolution of the American Hospital." In *The American General Hospital: Communities and Social Contexts*, edited by Diana Elizabeth Long and Janet Golden, 3-17. Ithaca: Cornell University Press, 1989.

Rosenkrantz, Barbara Gutman. "Cart before Horse: Theory, Practice and Professional Image in American Public Health, 1870-1920." *Journal of the History of Medicine and Allied Sciences* 29 (1974): 55-73.

Rosner, David. *A Once Charitable Enterprise: Hospitals and Health Care in Brooklyn and New York, 1885-1915.* Cambridge: Cambridge University Press, 1982.

Rossiter, Margaret W. *Women Scientists in America: Struggles and Strategies to 1940.* Baltimore: Johns Hopkins University Press, 1982.

Ruffman, Alan and Colin D. Howell, eds. *Ground Zero: A Reassessment of the 1917 Explosion in Halifax Harbour.* Halifax: Nimbus, 1994.

Rutty, Christopher J. "The Middle Class Plague: Epidemic Polio and the Canadian State, 1936-37." *Canadian Bulletin of Medical History* 13 (1996): 277-314.

– "'Do Something! ... Do Anything!' Poliomyelitis in Canada 1927-1962." PhD thesis. University of Toronto, 1995.

Sacks, Karen Brodkin. *Caring by the Hour: Women, Work, and Organizing at Duke Medical Center.* Urbana: University of Illinois Press, 1988.

Samson, Daniel. "Introduction: Situating the Rural in Atlantic Canada." In *Contested Countryside: Rural Workers and Modern Society in Atlantic Canada, 1800-1950*, edited by Daniel Samson, 1-33. Fredericton: Acadiensis Press, 1994.

Sandelowski, Margarete. *Devices and Desires: Gender, Technology, and American Nursing.* Chapel Hill: University of North Carolina Press, 2000.

Sangster, Joan. *Earning Respect: The Lives of Working Women in Small-Town Ontario, 1920–1960.* Toronto: University of Toronto Press, 1995.

"Saskatchewan Hospitals Protest Student Nurse Salary Order." *Canadian Nurse* 23 (September 1946): 41.

Scammell, H.L. "A Brief History of Medicine in Nova Scotia (Part 2)." *Dalhousie Medical Journal* (January 1966): 79–83.

– "A Brief History of Medicine in Nova Scotia (Part 3)." *Dalhousie Medical Journal* (May 1966): 93–5.

Scarselletta, Mario. "The Infamous 'Lab Error': Education, Skill and Quality in Medical Technicians' Work." In *Between Craft and Science: Technical Work in US Settings*, edited by Stephen R. Barley and Julian E. Orr, 187–209.

Scott, Donald M. "The Profession That Vanished: Public Lecturing in Mid-Nineteenth-Century America." In *Professions and Professional Ideologies in America*, edited by Gerald Geison, 12–28. Chapel Hill: University of North Carolina Press, 1983.

Sebas, Mary B. "Developing a Collaborative Practice Agreement for the Primary Care Setting." *Nurse Practitioner* 19, no. 3 (1994): 49–51.

75 Years of Caring: St Joseph's Hospital (Saint John, NB).

Shapin, Steven. *A Social History of Truth: Civility and Science in Seventeenth-Century England.* Chicago: University of Chicago Press, 1994.

– "The Invisible Technician." *American Scientist* 77 (1989): 554–63.

Shearer, A.R., ed. "Canadian Society of Laboratory Technologists: A Chronology 1937–1980." Unpublished manuscript. CSLT National Office, Hamilton, Ontario, [1983].

Siekawwitch, Eleanore (General Hospital, Moose Jaw, Saskatchewan). Letter to the Editor. *Canadian Journal of Medical Technology* 2 (1940): 146.

Simmons, Christina. "'Helping the Poorer Sisters': The Women of the Jost Mission, Halifax, 1905–1945." *Acadiensis* 14 (Autumn 1984): 3–27.

Smith, Susan. "White Nurses, Black Midwives, and Public Health in Mississippi, 1920–1950." In *Women and Health in America*, edited by Judith Walzer Leavitt, 444–58. Madison: University of Wisconsin Press, 1999.

Smyth, Elizabeth, Sandra Acker, Paula Bourne, and Alison Prentice, eds. *Challenging Professions: Historical and Contemporary Perspectives on Women's Professional Work.* Toronto: University of Toronto Press, 1999.

Soothill, Keith, Lesley Mackay, Christine Webb, eds. *Interprofessional Relations in Health Care.* London: Edward Arnold, 1995.

Sproule-Jones, Megan. "Crusading for the Forgotten: Dr Peter Bryce, Public Health, and Prairie Native Residential Schools." *Canadian Bulletin of Medical History* 13 (1996): 199–224.

Star, Susan Leigh. "Sociology of the Invisible." In *Social Organization and Social Process*, edited by David R. Maines. New York: Aldine de Gruyter, 1991.

Stevens, Rosemary. *In Sickness and in Wealth: American Hospitals in the Twentieth Century.* New York: Basic Books, 1989.

Strauss, Anselm. "Structure and the Ideology of the Nursing Profession." In *Professions, Work and Careers*, edited by Anselm Strauss, 24–67. New Brunswick: Transaction Books, 1975. Originally published in *The Nursing Profession*, edited by Fred Davis. New York: Wiley, 1966.

Strauss, Anselm, with Rue Bucher. "Professions in Process." In *Professions, Work and Careers*, edited by Anselm Strauss 9–23. New Brunswick: Transaction Books, 1975.

Sykes, R., and A.S. Sethi. "The Labour Movement in Health Care: Canada." In *Industrial Relations and Health Services*, edited by Amarjit Singh Sethi and Stuart J. Dimmock, 42–53. London: Croon Helm, 1982.

Taylor, Malcolm G. *Health Insurance and Canadian Public Policy: The Seven Decisions that Created the Canadian Health Insurance System and Their Outcomes*. Montreal: McGill-Queen's University Press, 1978.

Thorngate, Alice. *That Far Horizon: The Medical Technology Program at the University of Wisconsin-Madison, 1925–1975*. Madison: A-R Editions, 1983.

Tillotson, Shirley. "We May All Soon Be 'First Class Men'": Gender and Skill in Canada's Early Twentieth Century Urban Telegraph Industry." *Labour/Le Travail* 27 (1991): 97–125.

– "The Operators Along the Coast: A Case Study of the Link Between Gender, Skilled Labour and Social Power, 1900–1930." *Acadiensis* 20 (1990): 72–88.

Todd, James Campbell, and Arthur Hawley Sanford. *Clinical Diagnosis by Laboratory Methods: A Working Manual of Clinical Pathology*. Philadelphia: W.B. Saunders Company, 1927.

Torstendahl, Rolf. "Introduction: Promotion and Strategies of Knowledge-Based Groups." In *The Formation of Professions: Knowledge, State and Strategy*, edited by Rolf Torstendahl and Michael Burrage, 1–10. London: Sage Publications, 1990.

Tuchman, Arleen. *Science, Medicine, and the State in Germany: The Case of Baden, 1815–1871*. New York: Oxford University Press, 1993.

Twohig, Peter L. "'Once a Therapist, Always a Therapist': The Early Career of Mary Black, Occupational Therapist." *Atlantis* 28, no. 1 (Fall/Winter 2003): 106–17.

– "The Rockefellers, The Cape Breton Island Health Unit and Public Health in Nova Scotia." *Royal Nova Scotia Historical Society Journal* 5 (2002): 122–33.

– "Public Health in Industrial Cape Breton, 1900–1930s." *Royal Nova Scotia Historical Society Journal* 4 (2001): 108–31.

– "Organizing the Bench: Medical Laboratory Workers in the Maritimes, 1900–1950." PhD thesis. Dalhousie University, Halifax, Nova Scotia, 1999.

– *Challenge and Change: A History of the Dalhousie School of Nursing 1949–1989*. Halifax: Fernwood Publishing and Dalhousie University, 1998.

Urquhart, M.C., ed. *Historical Statistics of Canada*. Toronto: Macmillan, 1965.
Verma, Dhirendra. "Medical Laboratory Technology Instruction in Nova Scotia." MA thesis. Saint Mary's University, Halifax, Nova Scotia, 1968.
Vincent, George E. *The Rockefeller Foundation: A Review for 1923*. New York: The Rockefeller Foundation, 1924.
Vogel, Morris J. *The Invention of the Modern Hospital: Boston, 1870–1930*. Chicago: University of Chicago Press, 1980.
Waite, P.B. *The Lives of Dalhousie University. Vol. 1, 1818–1925*. Montreal and Kingston: McGill-Queen's University Press, 1994.
– *The Lives of Dalhousie University. Vol. 2, 1925–1980*. Montreal and Kingston: McGill-Queen's University Press, 1998.
Walby, Sylvia. *Theorizing Patriarchy*. Oxford: Basil Blackwell, 1990.
– *Patriarchy at Work: Patriarchal and Capitalist Relations in Employment*. Cambridge: Polity Press, 1986.
Walker, A. "The Technician's Trials and Tribulations." *Canadian Journal of Medical Technology* 8, no. 2 (1946): 71.
Walkowitz, Judith R. *Prostitution and Victorian Society: Women, Class and the State*. Cambridge: Cambridge University Press, 1980.
Warner, John Harley. "The Rise and Fall of Professional Mystery." In *The Laboratory Revolution in Medicine*, edited by Andrew Cunningham and Percy Williams, 110–41. Cambridge: Cambridge University Press, 1992.
– "Ideals of Science and Their Discontents in Late Nineteenth-Century American Medicine." *Isis* 82 (1991): 454–78.
– "Science in Medicine." *Osiris*, 2d Series, 1 (1985): 37–58.
Weindling, Paul. "Scientific Elites and Laboratory Organisation in *fin de siècle* Paris and Berlin." In *The Laboratory Revolution in Medicine*, edited by Andrew Cunningham and Percy Williams, 170–88. Cambridge: Cambridge University Press, 1992.
Weir, George. *Survey of Nursing Education in Canada*. Toronto: University of Toronto Press, 1932.
Whalley, Peter R., and Stephen R. Barley. "Technical Work in the Division of Labor: Stalking the Wily Anomaly." In *Between Craft and Science: Technical Work in US Settings*, edited by Stephen R. Barley and Julian E. Orr, 23–52. Ithaca: Cornell University Press, 1997.
White, Jerry P. *Hospital Strike: Women, Unions, and Public Sector Conflict*. Toronto: Thompson Educational Publishing, 1990.
Witz, Ann. *Professions and Patriarchy*. London: Routledge, 1992.
Wright, Anne. "Administration in Small Hospitals." *Canadian Nurse* 37 (April 1941): 230.

Index

Abramson, Harry L., 28, 30, 34, 50, 161; and problems in Bureau of Laboratories, 52–3
Acadia University, 95, 124
accuracy of laboratory tests, 43, 54–5
Adams, Tracey, 11, 138, 163–4
Agnew, G. Harvey, 105, 129
Allan, Dorothy, 62
allied health workers, 6, 140
American Medical Association (AMA), supervision of laboratories, 55
American Society for Medical Technologists, 140
American Society of Clinical Pathologists (ASCP), 88
anesthesia, 62, 85
Antigonish, NS, 58
antitoxin, 16, 48
Arnold, Grace, 111
autopsies, 28

Babkin, Boris, 75
bacteriology, 19, 34, 38, 89
Baker, Albert, 69
biochemistry, 19, 76–9
Blanchard Fraser Memorial Hospital (Sussex, NB), 125
blood bank, 7; blood chemistry, 50, 52, 57; blood counts, 51, 62; blood supply, 6; blood transfusions, 49–50
Branch, Arnold, 94; on salaries in New Brunswick, 120
Brison, Eliza Perley, 68
Bruce, Robert, 9
Bryce, P.H., 44
Bureau of Laboratories (Saint John, NB), 6, 15, 19, 27–30, 35, 52, 94; analyses conducted, 27, 28; need for more equipment 28, 50, 52–3; need for more space, 28, 50, 52–3; need for more staff, 52; and reportable diseases, 30; and salaries, 120; and training, 60, 66

Campbellton, NB, 29
Canadian Hospital Council, 97, 120
Canadian Journal of Medical Technology, 98–9
Canadian Life Insurance Officers Association, 116
Canadian Medical Association (CMA), 42, 112–14; and milk, 42; relations with CSLT 95, 112–14, 129
Canadian Red Cross Blood Transfusion Service, 126
Canadian Society of Laboratory Technologists (CSLT), 17, 73; code of ethics, 142–3; and history, 162; local academies, 100–1; membership, 87–8, 94–101, 130; membership categories, 95, 96; origins 86; provincial branches, 100–1; recruiting members, 133–6; registry, 86, 96, 133; relations with Canadian Medical Association 112–14; and World War II, 127–9
Canadian Society of Medical Laboratory Science, 7, 17, 165–6
Canadian Society of Radiological Technicians, 108
Canadian Tuberculosis Association, 15; and PEI Department of Health, 116–17, 119
cancer problem in Nova Scotia, 34, 50
Cann, Grace, 93
centrifuge, 53
cerebro-spinal fluid examinations, 27
Charlottetown, PEI, 116
Charlottetown Hospital (PEI), 89
Chase, Margaret, 13, 67

Chatham, NB, 29
chemistry, 7
Chisholm, William, 25
Chipman Memorial Hospital (NB), 85
Chipman, NB, 29
Clarke, C.K., 43
clerical work, 70, 74, 83–4
clinical judgment, 21, 34
clinical laboratory tests, 27, 30, 33
clinical medicine, 28, 35
Cockburn, Cynthia, 12
Colchester County Hospital (Truro, NS), 89, 90
combined x-ray and laboratory training, 64. *See also* laboratory workers education and Riddell, W.A.
Cunningham, Andrew, and Percy Williams, 8
Creelman, P.A., 116
Cruickshank, E.W.H., 76, 78
cryptosporidium, 6
cytology, 7

Dalhousie University, 25, 35, 123, 124; Department of Pathology and Bacteriology, 67; and medical science laboratories, 75–9; and nursing education, 24, 111; and women doctors, 67–8
Darling, George, 100
Dawson Memorial Hospital (Bridgewater, NS), 89, 119
Deadman, W.J., 105
Defries, Robert, 117
dental assistants, 11, 111
Department of Health (New Brunswick) established, 24
DeRicci, Sister Catherine, 86
Deutsch, Albert, 153
diagnostic services, 59; expansion of, 66. *See also* laboratories and x-ray

diagnostic technologies, 21
dietetics, 16
dietitians and multitasking, 91
diphtheria, 16, 31, 36; antitoxin, 40; free laboratory tests, 33; identified in laboratory, 40; laboratory-based diagnosis, 19, 38, 40–1; outbreak in Nova Scotia during World War I, 40–1; Schick test, 38, 40, 56; throat swabs, 27, 33, 38
Dock, Lavinia, on nurses in diagnostic service departments, 65
Dominion Coal Company, 58

Eager, W.H., 61, 64
Eastern Kings Memorial Hospital (Wolfville, NS), 85
ebola, 7
Elliot, Frank J., 95, 143; support for unionization, 147; on salaries in New Brunswick, 121
Empress of Ireland, 23
Etzioni, Amitai, 10
examinations, 112–13
expertise, 136–7

family wage for laboratory workers at Dalhousie, 77–8, 80–1
federal government and venereal disease control, 44–5
fiscal restraints in Maritimes, 35
FitzGerald, John, 117
Freidson, Eliot, 10, 158

Geison, Gerald, 9
gender, 161–4; and auxiliary hospital workers, 12; and demarcation strategies, 10–11; and family wage, 77–8, 80–1; and hospital division of

labour, 10–11; and laboratory boys, 75–9; and male physicians, 12; and multitasking, 11; and professional women, 139–42; and professional work, 11, 13. *See also* entries under laboratory workers
Gerard, Sister Catherine, 93
gonorrhea, 43, 44
Gowanloch, Louise, 124–5

Halifax and laboratory tests, 23, 66
Halifax Infirmary, 93
Halifax Ladies' College, 124
Halifax Medical College, 20, 21, 22; reintegration into Dalhousie University, 24
Halliday, Andrew, 20–1
Hamilton, Annie, 67
Hamilton General Hospital, 86
Harris, D. Fraser, 5–6
Harvey, Kathleen, 122
Hattie, W.H., 20, 31, 33, 41
health care system, idiosyncratic, 14
health care restructuring, 3
health care workers, social organization of, 3
health information, 33
health professional education, variations in, 14. *See also* laboratory workers education
health services, uneven development of, 155
hematology, 7, 89
hepatitis C, 6
Hines, Gertrude, 66
histology, 7, 31, 69
HIV, 6
Hopgood, Ella, 68
hospital history, and omission of laboratories, 8; need for case studies, 13

hospitals, addition of new workers, 93; development in United States, 58; development in Ontario, 58, development in Maritimes, 58; and informal division of labour, 92; multitasking workers, 154–5, 157, 161; wages and salaries for selected positions, 120–2
Hotel Dieu Hospital (Campbellton, NB), 29
Hotel Dieu Hospital (Chatham, NB), 8
Howell, Joel, 14
Hughes, Everett, 10

immunology, 19
infant mortality rate in Halifax, 42
institutional capture of laboratory workers, 138
insulin treatment, impact on laboratory, 52
interprofessional collaboration, 3–4

Junior League women working in laboratories, 129

Kahn, R.L., 5
Kahn tests, 50; Kahn tests for Prince Edward Island, 51
Keeping, Benjamin C., 116
Kemp, Ileen, 99–100; and unionization, 148
Kenney, W.W., 22, 25, 31, 64, 125
Kingston General Hospital (ON), 126
Kinnear, Mary, on Canadian hospitals, 59
Kirkpatrick, Elizabeth, 68
Kiwanis, 15
Koch, Robert, 38

Laberge, Louis, 19
laboratories, 16; development of, 19, 158–9; and medical history, 8; need for local study, 13–14, 160–1, 164–5; social impact of, 43; subspecialties, 7
laboratory boys, 75–9
laboratory diagnoses, 19–20, 56; diagnosis of diphtheria, 19, 56; diagnosis of tuberculosis, 19, 56; diagnosis of venereal disease, 56
laboratory directors, influence on CSLT membership, 96–7, 98
laboratory equipment, 30–1, 35, 37; cleaning and maintenance of, 46, 84
laboratory errors, 39, 54–5, 153–4
laboratory medicine, 8, 21, 28; relationship with bedside, 34–5
laboratory service in Nova Scotia, origins, 20
laboratory specimens, 31–2
laboratory standards, 54–5
laboratory technologists. See laboratory workers
laboratory tests, accuracy of, 43, 54–5; clinical tests 21, 30, 32, 56; clinical tests in Saint John, 72; and clinical judgment, 32, 34; diagnostic use, 19; expanding workload, 50, 53, 56, 72; false positives, 54–5; public health tests, 26, 27, 29, 32, 56; utility of, 19, 21
laboratory work, relationship with science and technology, 12
laboratory workers, 7, 72; advertisements for, 90–1; ambiguous roles, 82–3; connections with Dalhousie, 69–70; dedicated staff, 104; definition of, 136–47; and diagnosis, 143; different labels, 82–3; educational background, 88; expertise, 137; idealized portrait, 143–4; invisibility of, 7, 9, 162; marriage, 73, 78; marriage and attrition, 131; men, 75; mobility, 125–7; multiple identities, 154; multitasking, 74, 84, 89–90, 141–2, 154–7, 161; recruitment, 133–6; recruitment in Maritimes, 123–4; recruitment of men, 132; relationships with other workers, 14–15, 157; and respectability, 125–6; shortage 103, 106–7; shortage during World War II, 129; shortage during 1950s, 130–1; and student labour, 102–3; and unionization, 147–50; women, 16, 132; types, 142; volunteers, 70, 71
laboratory workers' education, 60, 101–12; apprenticeship culture, 146; approval by AMA, 105; approval by CMA, 105, 113; approved schools, 105–6; approved schools in Maritimes, 106; combined x-ray and laboratory program, 64, 108–9; continuing education, 82, 102; curriculum, 102; education on the job, 70–1, 82; entrance standards, 107; general technicians, 105; need to expand training programs, 106–7; at Pathological Institute, 61; role of commercial laboratories, 105; standardization, 104–5; specialty technicians, 105, 130; student salaries, 103; syllabus of studies, 102; training New Brunswick workers, 55; variability

in training settings, 103; varying training period, 61, 102; university education, 59, 70, 72, 109–10, 112, 115; university education at Dalhousie University, 110, 127
laboratory workers, relations with physicians, 111; insubordination, 111; institutional capture 138; medical dominance, 138; concern with behaviour, 143–4; personal scrutiny, 146–7
laboratory workers' wages and salaries, 73, 74, 119–22, 132; need for regular increases in Nova Scotia, 55; salary standardization 119; variation, 120–1, 122; in New Brunswick, 120–2; impact of university degrees, 125
labour process in hospitals, 92
labour turnover, 80
Larkin, Gerald, 4, 12, 158
Latour, Bruno, 14
Laval University, 110
licensed practical nurses, 3
Lindsay, M.A., 22–3, 25, 26
"local girls," 71, 120, 122
local government, 14. See also municipalities
Low, Margaret, 5, 37, 50, 82–3; appointed to do venereal disease testing, 45; public health work, 68–9

MacKay, Katherine, 67
MacKeen, R.A.H., 102, 124
MacKenzie, A.S., 5, 25, 26, 75, 77, 78
MacKenzie, D.J., 5, 20, 26–7, 35, 49, 68–9, 116; on training laboratory workers, 60; attitude toward CSLT, 96

MacKenzie, Eliza, 67
MacKenzie, Jemima, 67
MacKinnon, Clara, 63
MacLennan, Electa, 111
MacPherson, Greta, 62
Maritime Medical News and laboratory, 21
Massachusetts-Halifax Health Commission, 5, 53, 159; addition of laboratory assistant, 59–60, 66
McDonald, Flora K., 62
McKay, Ian, 14
medical bacteriology, 38
medical care, three sites of, 8
medical dominance of laboratory workers, 138
medical education, Dalhousie University, 24, 26, 37–8. *See also* Pathological Institute and Dalhousie University
medical laboratory technologists, 7. *See also* laboratory workers
medical records, 16
medical science, 19, 21, 34; medical science laboratories at Dalhousie University, 75–9
medical technology, reification of, 8
medicine, art of, 21
medico-legal laboratory tests, 27
meningitis, 31
Merry, Marion, 116–17
microbiology, 7, 19
microscope, 31
microtome, 31
Middleton Hospital (NS), 122
midwifery, 3
milk purity, 20, 42–3
milk testing, 28, 31, 35, 56; in Halifax, 66; in New Brunswick, 50
Mirimachi Hospital (Newcastle, NB), 84, 126

Mitchinson, Wendy, on need for Canadian analyses of health and medicine, 13
Moncton Hospital (NB), 29, 35, 55, 95, 125, 126
Montgomery, David, on labour turnover, 80
Montreal General Hospital, 126
Morse, Harry D., appointed to laboratory staff, 66
Mountain Sanatorium (Hamilton, ON), 102
multidisciplinary clinical teams, 3
multitasking, 4, 11, 153–4; and laboratory workers, 74, 80, 84, 89–90, 141–2, 154–6, 161; and nurses 62–3, 85–6, 161
municipalities 14, 35; responsible for health, 13
Murray, Florence, 5, 67
Murray, Foster, 66
Murray, George (premier of Nova Scotia), 22
Murray, L.M., 22, 23
Munro, Blanche, 67

national standards, 112–13
New Brunswick Civil Service Act, 119
New Brunswick Department of Health, 29
News Bulletin (CSLT), 99
New Waterford General Hospital (NS), 85
Nicholls, A.G., 23, 50, 61, 64; appointed, 31; on labour turnover, 80; on training laboratory workers, 60; unhappy with staffing levels, 68; role in recruiting workers, 69, 124
Nova Scotia, appointment of a provincial bacteriologist, 20; and laboratory tests, 23

nurses, 92; integral to hospital development, 60–1, 65; flexible labour pool, 11, 61, 62; and laboratory work, 62–3, 93; and multitasking, 16, 62–3, 60–5, 85–6, 89–90, 161; occupational boundaries, 93; student salaries, 103; and technical education, 64–5, 104; and technology, 65, 92
Nova Scotia Department of Health, 40
Nova Scotia Sanatorium (Kentville, NS), 95
nursing assistants, 92
nurse practitioners, 3

occupational boundaries, 4, 11, 65, 83–6, 90, 92, 158
occupational closure, 13
occupational groups, 10; heterogeneity of, 87
occupational therapists, 93
O'Donnell, Florence, 67
Olding, Clara May, 67
oral history, 17–18
Ottawa Civic Hospital (ON), 126
Owen Sound General and Marine Hospital (ON), 104

Park, William H., 28
Pasteur, Louis, 9
pasteurization, 42
Pathological and Bacteriological Laboratory Assistants Association (UK), 140
Pathological Institute (Halifax), 5, 15, 19, 22–7, 35, 116; administration of, 26; expansion, 24; expansion of public health section, 27; opened in 1914, 22; relations with Dalhousie, 23, 24–5; relations with Nova Scotia Department of Public Health, 25–6; staff in pathology section, 70; separation of pathology and public health work, 83; training, 60, 66; venereal disease testing, 45
pathology specimens, 31
pharmacy, 16
pharmacists and multitasking, 84, 91
philanthropic foundations, 35. *See also* Rockefeller Foundation
Philp, Martha, 67
physical therapists, 93
physicians use of laboratories, 30
physiology, 76
polio, 47–9; antipolio serum, 48; nasal spray, 49; preparation of convalescent serum, 47–9, doubts about efficacy of convalescent serum, 49
postal regulations and shipping specimens, 32
practical nurses, 92
Prince Edward Island, appoints laboratory worker, 60, 150
Prince Edward Island Department of Health, 116–17, 119
profession, 10; as an analytical category, 12, 139; system of professions, 154
professional boundaries, 3, 4; professional formation, 10
professional identity, 4; hallmarks of, 87, 114
professional women, 139–42; exclusivity, 141
professions, historiography of, 139
provincial governments and development of health care infrastructure, 14, 35; limited fiscal resources, 20
pseudonyms, use of, 18

public health, development of, 15; link with laboratories, 19–20, 21, 26; and laboratory tests, 27, 33, 37–8; public health nurses, 24
Purdy, Evelyn, 63

radiographers, 11, 12. *See also* x-ray technicians
Red Cross, 16, 116, 126
Regina College and laboratory education, 109
Regina General Hospital and unionization, 147
registration of laboratory workers; with American Society of Clinical Pathologists (ASCP), 88; with CSLT, 94–6; CSLT registration exam, 94; "registered technologist" (RT) designation, 96
regional framework of analysis, 13–14
Reiser, Stanley on organization of health services, 59
reportable diseases, 30
Reid, A.P., 21, 33, 43
respiratory technologists, 93
Reverby, Susan, 60; on labour process, 92; on nursing education, 86
Rice, Grace, 67
Riddell, W.A., 108–9, 115
Roberts, William F., 27, 30, 42
Rockefeller Foundation, 16, 26, 118
Rockefeller Institute for Medical Research, 47
Rockefeller Hospital (New York), 47
Rosner, David, 14
Royal Canadian Navy Hospital (Halifax, NS), 126
Royal Columbian Hospital (New Westminster, BC), 126
Royal Victoria Hospital (Montreal), 125

St Catharines General Hospital (ON), 103
Saint John General Hospital, 28, 29, 53; and insulin treatment, 52
St Joseph's Hospital (Glace Bay, NS), 58
St Joseph's Hospital (Hamilton, ON), 103
St Joseph's Hospital (Victoria, BC), 104
St Martha's Hospital (Antigonish, NS), 60; and laboratory training, 101–2
St Mary's Hospital (Inverness, NS), 89
St Thomas Hospital (London, UK), 125–6
Sandelowski, Margarete, 60, 65, 85
Saskatchewan Hospital Association, opposition to student salaries, 103
scientific medicine, 8, 35, 159
semiprofession, 10, 133
serology tests, free in New Brunswick, 51
serums, 38
service ideal, 142
service organizations, 15
severe acute respiratory syndrome (SARS), 7
sex segregation, 12
Shapin, Stephen, 9
Sisters of Charity, 59
Sisters of Saint Martha, 58
skill, 6, 57, 166; and laboratory work, 137–8, 145–6; and service, 137, 145; technical skill, 137, 156
smallpox, 16
Smith, Ralph P., 105, 146; on laboratory workers' salaries, 121
specialization, 80, 107–8
specimens, collection of; 32; shipping of, 31–2
speech language therapists, 93
Spencer, Minnie, 67

sputum samples 33, testing procedure 38–9. *See also* tuberculosis
staff, expanding need for, 37, 50; levels in Halifax 68; retention in Halifax, 55
staining samples, 38, 39
Stanley, Carleton W., 78
Stevens, Rosemary, on nurses, 65
Stevenson, Esther, 117–18
Strauss, Anselm, on heterogeneity of occupational groups, 87
students, paid during training, 103
Summerside, PEI, 116
syphilis, 44; free tests in Halifax, 44; importance of testing accuracy, 43; Kahn tests, 50; Kahn tests for Prince Edward Island 51; Wassermann test, 38, 44, 50, 56; modification of Wassermann test in Halifax, 54; Wassermann test procedure, 45–7; Wassermann tests for Prince Edward Island, 51

Technical College of Nova Scotia, 22
technical work, 6, 7; separated from diagnostic work, 12, 79
technologists, x-ray, 7
Thorndike Memorial Laboratory (Boston, MA), 125
Thorngate, Alice, on multiple roles of laboratory workers, 84
Thurott, Bessie, 67
tissue examinations, 22, 28, 34, 42, 50, 69
toxoid, 16
tuberculosis, 7, 16, 31, 36; sputum examinations, 33, 37; sputum samples, 38
tumour examinations, 22

typhoid fever, 16, 31, 36; and blood specimens, 37; free tests, 33; laboratory-based diagnosis, 19; and sewage disposal, 41; and water purity, 41; Widal test, 38; Widal test procedure, 38–9

urine tests, 33, 62, 89; procedure for urinalyses, 51–2
unions, 4; and laboratory workers, 147–50; and nurses, 148
university education for laboratory workers. *See* laboratory workers' education
University of Manitoba, 111
University of Saskatchewan, 110
University of Western Ontario, 110

vaccination, 16
vaccines, 48; vaccine preparation, 38; in New Brunswick, 28; in Nova Scotia, 47
venereal disease, 16, 36, 43–7; and federal health department, 44; laboratory testing grows, 27, 45; provincial legislation, 44; in New Brunswick, 50, 52; treatment centres in Nova Scotia, 44; "VD problem," 16. *See also* syphilis, Kahn, Wassermann
Victoria General Hospital 4, 22, 83, 85; cancer clinic, 50; and laboratory tests, 23; management of laboratory, 24; and training x-ray and laboratory workers, 64; x-ray department, 61
Victoria Public Hospital (Fredericton, NB), 55, 124
vital statistics, 33

Vogel, Morris, 14
voluntary agencies, 15–16

Walkerton, ON, 6
Warner, John Harley, 8
Warwick, William, 8, 51
Wassermann test, 38, 50, 56; diagnostic value, 47; false positives, 47; modification of Wassermann test in Halifax, 54; Wassermann testing procedure, 45–7; Wassermann tests for Prince Edward Island, 51

water testing, 20, 27, 28, 31, 35, 41, 56
Weir, George, 63, 86
West Nile Virus, 7
white-collar work, 140–1
Widal tests, 33, 38–9, 56
Witz, Anne, 10
Wodehouse, Robert E., 117
women and technology, 12
women physicians in Nova Scotia, 67–8; "feminine specialties," 68
Women's Institutes, 15
Woodstock, NB, 29

Woolgar, Steve, 14

x-rays and clinical judgment, 34; and diagnosis, 12; x-ray departments, 16; x-ray technology, 10; separation of manual and diagnostic work, 61; x-ray work at Victoria General Hospital, 61

Yarmouth, 93
Young, E. Gordon, 76, 77, 78, 79, 80